AT THE
SERVICE QUALITY
FRONTIER

Also available from Quality Press

Quality Service Pays
Henry L. Lefevre

Deming's 14 Points Applied to Services
A. C. Rosander

Implementing Quality with a Customer Focus
David N. Griffiths

Quality Dynamics for the Service Industry
W. F. Drewes

Quality Service—Pure and Simple
Ronald W. Butterfield

The Customer Is King!
R. Lee Harris

The Quality Revolution and Health Care
M. Daniel Sloan and Michael Chmel, M.D.

Quality Assurance in the Hospitality Industry
Stephen S. J. Hall

Quality Management in Financial Services
Charles A. Aubrey II

To request a complimentary catalog of publications, call 800-248-1946.

AT THE SERVICE QUALITY FRONTIER

A Handbook for Managers, Consultants, and Other Pioneers

Mary M. LoSardo
Norma M. Rossi

ASQC Quality Press
Milwaukee, Wisconsin

AT THE SERVICE QUALITY FRONTIER
Mary M. LoSardo, Norma M. Rossi

Library of Congress Cataloging-in-Publication Data
LoSardo, Mary M.
    At the service quality frontier: a handbook for managers,
consultants, and other pioneers / Mary M. LoSardo, Norma M. Rossi.
        p.   cm.
    Includes bibliographical references.
    ISBN 0-87389-209-7
    1. Service industries—Quality control. 2. Customer service.
    3. Service industries—Management. I. Rossi, Norma M. II. Title.
    HD9980.5.L67   1993
    658.5'62—dc20   92-33879
                                                        CIP

10 9 8 7 6 5 4 3 2 1

ISBN 0-87389-209-7

Acquisitions Editor: Susan Westergard
Production Editor: Mary Beth Nilles
Marketing Assistant: Robert N. Platt
Set in Avant Garde and Galliard by Montgomery Media, Inc.
Cover design by Montgomery Media, Inc. Printed and bound by BookCrafters.

For a free copy of the ASQC Quality Press Publications Catalog,
including ASQC membership information, call 800-248-1946.

Printed in the United States of America

  Printed on acid-free recycled paper

ASQC
Quality Press
611 East Wisconsin Avenue
Milwaukee, Wisconsin 53202

For my parents, Laura and Attlio Rossi

—NMR

In loving memory of my aunt and uncle,
Gertrude Mullins and John P. Mullins

—MML

C O N T E N T S

P R E F A C E

All pioneers face many challenges and obstacles as they push against frontier boundaries. Some are unable to deal with them and fall by the wayside. Others, in the face of seemingly insurmountable barriers, still manage to push ahead and reach their goal because they are drawn by a clear vision and have the support and companionship of fellow travelers who believe in what they are trying to accomplish.

Any success we have had in pushing forth the boundaries of our frontier must be shared with those who provided the guidance, support, and belief we needed to help us overcome the obstacles in our path. Our thanks go to John Creedon, who provided the vision and never wavered in his belief that it was attainable; to John Falzon, who had the courage to help us translate the vision into reality; to Loretta Francone, who remembered the details that kept us on the right path; and to the late Neal Campbell, whose wealth of experience and unfailing humor helped us keep our eyes on the goal.

Mary LoSardo and Norma Rossi

INTRODUCTION

This book began in 1984 when, as internal consultants, we began to develop a quality improvement process for a major financial institution.

We were very lucky because the decision to begin came from the top, where quality was seen as the key to a competitive advantage, a way of distinguishing our company from the competition.

We were also fortunate because early on there was a recognition that quality is more a journey than a destination. Quality improvement was seen as more than reducing error counts. It represented a fundamental change in the way a firm does business. With such a philosophy, "quality" is not an add-on to one's job; it becomes one's job.

Because of this, continuous improvement is not something that can happen overnight, nor is it something that can be bought off a shelf. So, we were given the luxury of time up front to design and develop a process that would fit our organization and our industry.

As with any journey into unknown territory, we began by finding out as much as possible from others who had already begun to travel down this road. We reviewed everything we could get our hands on, we talked to many consultants, and we visited several companies.

All our research led to one shocking conclusion: While all those involved in quality improvement agreed on certain basic principles, the principles were only being practiced on the shop floor. Even those organizations that were successfully applying quality tools and techniques in their manufacturing operations had yet to extend them to the "white-collar" part of their companies. We could find no models of a successful quality process in the service industry.

This led to our first decision: to feed these findings back to our top managers—the people who would have to make quality improvement happen—and to ask them what our quality process must incorporate if it were to be successful.

Their input was crucial. First and foremost, they equated quality with the customer. Customer satisfaction became both a driver and a goal.

Second, they determined that this obsession with the customer had to be shared by each and every employee. All employees had to identify their customer;

but understanding wasn't enough. An infrastructure had to be developed that would encourage employees to identify opportunities for improving customer satisfaction and to respond to them. Moreover, employees need to participate actively in the decision-making process in their work areas.

For many service organizations, translating these ideas about customer orientation and employee involvement into action is indeed revolutionary. For years, services have organized themselves around the models of the Roman legion or the automotive assembly line. The norms are clear but constrained by rigid lines of authority, with one-way, downward communication of goals, objectives, and directions. Obedience to authority was rewarded. Work processes were broken into ever-smaller, more specialized functions in the interest of efficiency.

The primacy that our top managers were giving to the customer and the employee signaled a break with these traditional beliefs. Obviously, they were asking the other members of the organization to alter the way they worked, to follow new rules of the game.

In any changing culture, anyone who has been successful under the old rules can find it difficult to accept or follow the new ones. We met this challenge in two ways. First, we translated the principles that would underlie our quality improvement efforts into three broad objectives that any department or division, any team or individual, could customize into specific goals and quality improvement projects: (1) improving customer satisfaction, (2) achieving higher levels of performance, and (3) eliminating "extra processing"—the additional resources consumed when things aren't "done right the first time."

Our second decision was to continue to develop "buy-in" to the process by letting the members of our organization use their own life experiences as customers to arrive at the same quality principles. As with top management, we began our organizational change efforts with what we called a "past opportunity" exercise. Here, individuals were asked to identify a time when, as customers, they did not receive a good product or service. What happened? What would they have liked to have happened?

By sharing these experiences in small group sessions, both managers and employees were able to see certain basic themes arise. Together, they could begin to look at their own workplace and use some basic tools and techniques to identify opportunities for improvement—opportunities that would fall into the three broad-based goals described above.

In this book, we have set out to describe the process we went through. We describe the thought processes that led us to our customer-based philosophy and discuss how we applied it. We cover the breakthrough research we discovered was being conducted in the marketing field and the practical way it can be applied. In short, a manager or consultant could look at this as a cookbook for designing a quality improvement effort.

However, this work can serve as much more. Along with the specific descriptions of what can be done, we have included many of the "past opportunities" we have come across in the course of our consulting. These experiences have been retold in the form of stories. They are situations that we have lived through ourselves or that our participants have shared with us. None of them is fiction. We believe that many readers can "see" themselves in similar (sometimes identical) circumstances and can relate directly to them. For this reason, we believe that the stories shorten the distance between knowing the theory of service quality and making that theory a reality.

An example of the kind of story you can find throughout the book follows. It highlights one of the truths we have discovered about managing service quality over the years; that is, that actually creating service quality for the customer is much more difficult that talking about it, planning for it, or even writing about it, even when you consider yourself a "quality expert."

ْ ْ

Quality Is as Quality Does

Before leaving for home the evening before, Ron checked with his administrative assistant. "Is everything OK for tomorrow's presentation, Marie?" "Sure. What could go wrong? We've done this so many times before. See you in the morning," she replied cheerfully.

But the minute he stepped off the elevator the next morning, he had a premonition that things were anything but OK. He was right. Marie came running into his office, and without waiting for him to remove his coat, she said, "A bunch of your guests are already here! They've been to my office asking for you. I don't believe people can come to a meeting so early!"

"I don't understand it," said Ron, glancing at his watch. "They aren't supposed to be here for another hour. The presentation doesn't start until nine-thirty—it isn't even eight-thirty. Is Jackie in yet?"

"I think so. I'll go check."

Ron nodded and headed for a quick stop at the coffee machine.

Ron and Jackie were part of their company's corporate quality staff. At their boss' request, they had designed a whole day's formal presentation that focused on the history of their company's quality improvement process. Whenever people from other organizations called asking for this information, the staff extended an invitation to spend the day hearing about quality. As a courtesy, guests received free coffee, sodas, snacks, and lunch.

From the start, these "Quality Days" were something that both Ron and Jackie looked forward to. They enjoyed sharing important information with their counterparts in other companies and governmental agencies, and they were gaining something as well. The people who came were sincerely trying to create better service quality and freely shared their ideas and experiences.

However, as their boss so often reminded them, if you invite people to a day that celebrates your company's efforts at service quality, it follows that it should be carefully planned to reflect that commitment. The unexpectedly early start of today's presentation did not inspire Ron with confidence in that commitment.

He hurried to the conference room, carrying a slide projector and the slides that would be used for the presentation. He didn't enjoy setting up the room in front of guests, but he could do little but smile, introduce himself, shake their hands, and get on with it. He noticed that more than half of the dozen guests they expected today were already there. He was immediately ashamed at the thoughts that came to him about guests who come too early or stay too late.

Ron completed setting up the projector but realized that the screen had not been moved down from the ceiling the previous day, as had been promised. He ran out of the conference room in a futile search for someone who knew which button to press. Why didn't people do the things they promised, he thought.

When he got back to the conference room, Jackie was there, giving out the folders that contained their handouts. "Thank heavens you're here," he whispered, "I don't know how to get the screen down—do you?"

"Sorry, I don't, but that's not all we have to worry about. I've still got to put the videotape in the VCR—which is locked in that cabinet over there. Do you have the key?"

"Sorry, I don't." They looked at each other without knowing whether to laugh or cry.

"I thought the conference room administrator was supposed to do that before she went home yesterday," said Jackie softly. "And by the way, the coffee, tea, and muffins aren't here yet. This is ridiculous—the guests are early, but the coffee is late."

At that moment, one of the guests, a senior vice president from another company, looked at his watch and asked "I guess we'll get started soon?" Jackie replied sweetly, "Well it's still rather early and not everyone is here yet." The vice president responded sourly, "It's not that early—it's nearly eight-thirty." Boy, does this guy have an attitude problem, thought Jackie.

She hurriedly went back to Marie's office. "Please call the building services people to check where the coffee and muffins are. They should have arrived already."

"I did. They told me they never got my memo requesting breakfast service. Here's my copy," she said, waving a piece of paper like a triumphant banner. "I told them that we needed it as soon as possible, so they're coming, but they can't be there until nine-thirty or ten o'clock."

"Can't be here or won't be here. Isn't it just like them to say they never got the memo," said Jackie angrily. "It's just an excuse." She sighed in resignation. "My getting upset is not going to produce the coffee. I guess we'll have to ask for written confirmation from now on."

When she entered the conference room again, the screen was down and the cabinet was open. She wondered how Ron had performed that minor miracle. Even though it was only nine o'clock, he had already begun his part of the presentation. Jackie noticed that all the seats were filled. Boy, this crowd really likes waking up early, she thought.

Just then, a man entered the room, looked around hesitantly for his name on a place card, crossed in front of the beam projecting on the screen, and ended up sitting in the last seat in the room. He grinned sheepishly at Jackie. Ron and Jackie glanced at one another, and then Ron, notwithstanding that he was in the middle of his presentation, said to the man, "Good morning. Are you here for a presentation on quality?"

"Oh yes. My name is John Campbell, from Acme Truckers. I called about two months ago. I never got a letter of confirmation, but I decided the letter had gotten lost, so I came anyway. Sorry for being late. I thought your presentation started at nine-thirty."

Too stunned to say anything more, Ron replied, "Oh, fine. Welcome." He quickly returned to his presentation.

When Ron's part was over, he turned the show over to Jackie. While he sat listening to Jackie along with the guests, still trying to remember who had invited Mr. Campbell, Marie poked her head in the room calling him into the hall. "What's the matter now?" he asked as he joined her.

"I just checked with the executive dining room about lunch. Don't ask me why I did, I guess I was nervous after finding out about the coffee. They tell me they were never notified that we would have fourteen for lunch in one of the private rooms. And I can't find the memo requesting reservations anywhere. I think your other assistant, the one who got married just before she quit, forgot to send the memo," added Marie. "What should we do?"

She had barely gotten the question out when a porter arrived with the coffee and muffins. Before Ron could stop him, he barreled his way into the conference room, dragging a noisy cart in his wake. He interrupted Jackie, who promptly lost her train of thought. "I think this is an excellent time for a break," she said rather too quickly.

As their guests rose, she joined Ron and asked quietly, "What's wrong now?"

"Nothing really," he replied sarcastically. "It's ten o'clock, and I've just been told that the executive dining room never received our request for a luncheon in a private room. You also may have noticed that we now have an extra mouth to feed." Taking a deep breath, he continued, "I'm going back to the office to check the file and do battle with the reservations people if I have to—or beg on my hands and knees. Can you take over the next part of my presentation?"

"Sure. I've heard it enough times."

Back at his office, Ron phoned the dining unit supervisor, to no avail. Even begging wouldn't work. In an act of desperation, he told Marie to order a catered lunch from a local delicatessen and not to spare on the quantity or quality. "I'll authorize payment, whatever it is, so long as these people get fed."

"I'll start calling places right now," said Marie helpfully. "But I'm going to call the dining unit supervisor and ask her to put us on a waiting list. If they get a cancellation, they might still take us."

"Well, go ahead if you think it will do any good. But don't forget we now have one more person attending—it will be fifteen for lunch instead of fourteen. A guy showed up today who was listed in the file, but somehow we never sent him a confirmation letter. What's strange to me is that he showed up anyway."

When Ron arrived back at the conference room, Jackie was waiting outside. "I'm showing them the first video segment. What's the scoop on lunch?"

"Let's just say I've made an executive decision."

"Boy, are you mysterious today," said Jackie. "How did you manage the screen and cabinet?"

"It was easy. I cornered the conference room administrator the minute she showed up and threatened her life."

"Terrific way to win friends and influence people," said Jackie unhappily.

5

Ron took the criticism to heart but attempted a defense. "I was upset. By the way, while I was checking the file, I realized why these people showed up so early. We told them to in the confirmation letter! It says the meeting would start at *eight-thirty*, not *nine-thirty*, like usual."

"Oh, no! And I kept saying how early it was to begin the presentation! They must think we've lost our minds."

"I think they may be right," said Ron "After all, if my memory serves me right, neither of us checked the letter before it went out. I guess we expected too much of Fran before she quit. After all, she was planning a wedding, buying a house, and changing jobs all at once."

Jackie shot back, "I think she could have said something if it was too much for her."

The rest of the morning's presentation proceeded without further crises. Ron left the conference room early to get the details about the catered lunch from Marie. "Well, can we feed these people, or am I going to die of embarrassment?" he asked half jokingly.

"There's good news and good news," she reported. "You actually have a choice now. About an hour after I ordered the food, the dining room supervisor called and said there is a cancellation for lunch. You can go up to the executive dining room as planned or have your guests eat the food that'll be delivered in about twenty minutes."

"I think we should eat in the executive dining room—it's neater. What time are they expecting us?"

"In half an hour—about twelve-thirty. What should I do with all the food I ordered? There'll be enough turkey, roast beef, and ham for fifteen people. It's too late to cancel the order."

"Offer it to everyone on our staff and call around to your friends in the building. Just don't let it go to waste. About how much is the caterer charging us?"

"They estimated about two hundred fifty dollars."

Marie watched as Ron's face blanched. Through clenched teeth he said, "Well, I guess that's the price of doing things wrong. Wait until the boss sees this."

Marie reflected, "Yeah, and the two hundred fifty dollars doesn't include the cost of having our guests eat in the executive dining room."

"Thanks for reminding me," Ron said over his shoulder as he left.

Once they arrived in their private room, the luncheon posed no problems, except that both Ron and Jackie weren't very hungry. Privately, Jackie said, "Ron, I can't imagine what else could go wrong today."

Later that afternoon, she wished she had a better imagination. It was during her afternoon presentation that the next problem emerged.

As Jackie spoke about some research they had conducted on the nature of service quality, she advanced the slides showing the major points in her remarks. At a crucial point, there were no slides—just a blank screen.

Ron came to her rescue—sort of. "I guess we never replaced those slides after I returned from my trip to Kansas City. I apologize for this, folks, but right after our break you'll be seeing a video excerpt that reviews the research thoroughly. I see that our soda and snacks have arrived. Why don't we take a fifteen minute break."

As the room emptied, Jackie went to sit next to Ron, who was staring into space. "Thanks for calling a break. I didn't realize the slides weren't in the tray. A perfect ending to a perfect day." Ron didn't respond immediately. "Ron, are you OK?"

"Sure, I'm fine...I guess I'm all right...I don't know...definitely not. It's almost four o'clock. When they come back, we'll play the last video segment, let them ask their final questions, and I hope, get them out of here by four-thirty. I've had it for today."

"I can't wait until this day is over," sighed Jackie. "My nerves are totally frazzled. But Ron, tomorrow let's do a review on today's problems. We just can't let this happen again. I'm not sure I can survive another 'quality day' like this one."

❧ ❧

If anyone is wondering whether this episode is based on actual experience, we have to admit that it is. Meeting customer expectations is hard work, even for people who supposedly know better. Designing a service demands attention to detail, creativity, and sensitivity to both internal and external customers—but we are getting ahead of ourselves.

We believe that the episodes presented throughout this book can serve as the basis for training that helps an organization become more sensitive to what quality truly means in the service environment. Our hope is that these stories will be used, along with the questions for discussion that appear at the end of each chapter, as a means for teams and organizations to begin their own quality improvement efforts.

DEVELOPING A CUSTOMER ORIENTATION: THE ORGANIZATIONAL SELF-ASSESSMENT

Developing a successful customer-based company—one tuned in to the voice of the customer—begins with a basic self-assessment. As with any journey, unless you know your starting point, you have no idea of the direction in which to head or how much progress you're making.

This self-assessment must focus on an awareness of your organization's relationship to its customers. The approach doesn't require any special training or esoteric techniques. Instead, all you have to do is answer the four simple questions listed in Table 1.1:

Table 1.1 The Organizational Self-Assessment

1. What services do we produce and provide?
2. Who are the customers for each of these services?
3. How do we measure the effectiveness of the service we provide?
4. How do our *customers* measure the effectiveness of the service we provide?

While the questions are simple, the process of answering them leads to a great deal of soul-searching. Applying this method tests the assumptions that have guided our delivery of services.

So let's look at each of these questions in detail and develop a common understanding of what the focus must be. Because forewarning is forearming, we'll also take some side trips to examine the pitfalls that service providers commonly fall into when they engage in this process.

Question 1: What services do we produce and provide? At first glance, this question appears deceptively easy to answer. A bank director might respond with loans, checking accounts, savings accounts, and safe deposit boxes. A food retail store manager might specify the major classes of foods and nonedible products stocked on the shelves.

However, this initial reaction does not go far enough. While it's important to identify the "big" picture—the major outcomes customers expect—actual service does not spring forth in full bloom. Instead, service outcomes that customers expect result from the total of many successful, smaller service interactions that occur within an organization.

This brings us to an important principle: To be truly customer-driven, the process of discovering the answers to this first question must permeate the organization. It's not something that's just done by the executive office or by the marketing people or by the customer service department. Everyone has an obligation to understand what service he or she is responsible for producing.

If we keep answering the question of what service is produced and provided all the way throughout an organization, we find the individual service workers of our world. The answers move us logically from an imposing (but to most workers vague) view of service to one that is meaningful at all levels. The concept must be one that check-out clerks, telephone operators, programmers, and secretaries can relate to their everyday activities. Thus, **service** should be broadly defined as

an item produced or an activity engaged in that is used by, or needed by, the next person in the work process.

This definition recognizes that service is a product of many individual efforts. The service employee on the firing line is highly dependent on the contributions of support people who work behind the scenes. It is the contributions of these unseen, unsung heroes that provide value added to the finished product!

This discussion leads to yet another question. Why do we produce such services in the first place?

Question 2: Who are the customers for each of these services? When we think of customers, we usually think of those who are the ultimate users of our services—the people outside our organization whose dollars keep us in business. Our basic definition of customer is much broader, for we see a **customer** as

any individuals or group in the work process requiring your output to perform their work activities or because they are the ultimate users.

It doesn't matter if the person you provide service to is in the office down the hall or facing you across a cash register—that individual is your customer. But to assess our organization's customer awareness, we must first look at (1) the different types of customers we serve, and (2) the different types of customer relationships we engage in.

TYPES OF CUSTOMERS

Customers sometimes appear to be a strange breed—whimsical, demanding, illogical, and capricious. They are also our reason for being—without customers, we service providers wouldn't be around very long!

Customers come to us because they want or need the service we offer. We must realize there are two very different types of customers: those who come to us because they want to and those who come because they have to!

The customers who want to come to us are attracted by what we offer. They're free to choose from a host of other suppliers; however, for various reasons, they select us. Perhaps we offer reasonable rates, or our business hours are convenient for them, or our salespeople seem most knowledgeable. We actively compete with others in our industry for these customers. They're probably the ones that consume most of our attention whenever we plan a new service or evaluate our sales results.

The other customers, who come to us because they have no other choices available—they are our captives. In many cases, they are what we define as **internal customers:**

> **people or groups within the organization who are dependent on the work of—or input from—other areas within the organization.**

Without the activity or input of a "supplier" elsewhere in the company, an internal customer cannot perform his or her own work activities. For example, a payroll clerk needs accurate, timely information from the person who keeps attendance records, as well as an up-to-date payroll system. These supports enable the clerk to provide high quality service to his or her customers: the employees of the company.

In companies lacking a true customer focus, internal customers are often mistreated. They are taken for granted and their wishes are ignored because suppliers assume that the customers will always be there. Company policy doesn't "allow" them to seek new suppliers elsewhere.

On the other hand, sometimes even well-intentioned suppliers will provide their internal customers with inferior service. They mistakenly believe that they "know" the customers' wants and needs better than the customers do themselves. This is a common failing among suppliers who deal in complex or sophisticated services. In today's high-tech environment, we've all come across the technocrat who figuratively (and sometimes literally) pats us on the head while saying "Trust me, I know what's best for you." Such words are frequently the prelude to disaster.

However, captive customers need not be confined within our organizational bounds. Often, geographic or regulatory constraints will limit the choices available to external customers in the same way that company policy severely limits the choice of internal customers.

Let's look at a real-world example to see what can happen to a captive customer who's on the outside looking in for service.

è& è&

"What Do You Want From Me—I Only Work Here?"

"Francine! It's your turn to pick up the phone—your starting time was over 10 minutes ago."

"OK, OK. I'll get it," replied Francine. She reached over and switched on the phone while placing the headset on. "These customers are driving me crazy with their complaints."

"Then work hard and become a supervisor, Francine. I thank heaven every day that I don't have to deal directly with customers anymore."

Francine works for Hugh Power Services (HPS), a public utility company that serves a large metropolitan area. HPS recently published a new mission statement in the annual report, which states that the company "prides itself on serving the energy needs of our diverse community."

"HPS. How may I help you?" Francine turned her head to the back of the room checking to see whether the supervisor was listening into her line. No light by her number—that's good.

"Well, are you there or not?"

"Yes, I'm here," said a frail voice on the other end of the line.

"I need your account number. Please read it off your statement."

"Oh, what number is that? The one at the top or the one half way down the page?"

Oh sugar, thought Francine, another dummy. "The thirteen-digit number at the top beginning with 180000," she replied in exasperation.

"OK, I see it. It's 1800007655578."

"Please wait for the number to be verified," said Francine in her most official voice. She typed the number on her keyboard and waited for account information to appear on the video display terminal. This one probably hasn't paid her bill in months and is wondering why we are asking for the money, she thought. But when the information appeared on the screen, it showed that all payments were current. Francine was a little disappointed.

"Mrs. Mullens, my records show that you are paid up and that a meter reader has been to your home within the last three months—so what's your question?"

"I wasn't calling about the meter reader, although I would like to say that he is very nice and efficient. His name is Mike."

What did I do to deserve this punishment? thought Francine.

"Well, why did you call—if you don't mind my asking?"

"I don't mind. I'm calling because I don't understand one of the items on my latest bill. Let's see, it's right here, yes, it's called 'fuel surtax.' I notice that my bill is four dollars and ninety-four cents more because of it. Could you tell me what it is? Why am I paying almost five dollars more? It seems like a big increase all at once."

Francine exploded. "Listen, don't complain to me about rate increases—they don't show up in my paycheck. Didn't you see the notice that was sent out last month with your bill? It explained everything."

"Excuse me, but I didn't mean to upset you. All I want is an explanation of this fuel surtax thing."

"I already told you, the notice went out with last month's bills. Check around the house. I'm sure you'll find it."

"I really don't think I'll find it. I'm a senior citizen, and my son comes over every month to help me with the bills. I'm sure the notice was thrown out after he paid the bill."

"Then why don't you ask your son to explain things to you?"

"He didn't really know when I asked him, and he suggested I call you. Anyway, I'd really like to understand things for myself. I may be a senior, but I'm not stupid! You're there in customer service to help me and answer my questions, right?"

Francine chuckled. "Listen, you say you want customer service. Well, don't the lights go on every time you flip a switch? Try getting your electricity from some *other* company."

Francine paused, saving her strength for the next assault when she heard Mrs. Mullens hang up. Boy, she thought, what manners some people have—she didn't even say good-bye. What a way to make a living!

❧ ❧

Most of us, like Mrs. Mullens, cannot choose another utility company if we are unhappy with the one we've got. For instance, while we can choose long distance carriers, we're stuck with the local company that services our area. Moreover, if we're dissatisfied with local service, it may present a physical or financial hardship to switch to another grocery store or pharmacy.

Such "external captives" often suffer from the same slings and arrows that we inflict on our internal customers. They may experience service that ignores their needs, condescending treatment, and transactions completed at the convenience of the service provider rather than that of the customer.

Why are we making such a big deal of our captive customers? After all, we more or less own them—where else are they going to go! Well, things aren't that simple any more. First, if we're truly a customer-driven service provider, we will realize that customer satisfaction is the only true measure of quality. High-quality service means that we strive for the satisfaction of all our customers—external and internal.

Second, because no front-line service employee is an island, each is dependent on the back-room support of others in the work process chain of command. We can't expect to provide our external customers with high-quality service if our internal customers aren't receiving it.

Finally, captive customers who receive poor quality service eventually break their bonds. We've had many experiences with internal staffs who forgot that their purpose was to serve others within the organization. These units often became bloated and rigidly bureaucratic. Rather than help managers and service employees, they became stumbling blocks or even blockades.

What is the result? Given time, captive customers find ways to get around the blockade. They have to if they want to accomplish their mission. As soon as that happens, it clearly shows that the internal service is no longer needed. Captives manage 13 without it or do for themselves. The dissolution of the function—and disappearance of its staff—quickly follow.

Moreover, captive customers who are the end-users of a service, like Mrs. Mullens, also seek their revenge. It often takes the form of complaints to public utility commissions, consumer revolts against rate increases, and class-action lawsuits.

TYPES OF CUSTOMER RELATIONSHIPS

As we identify customers, it's important to realize that there is usually more than one type of customer relationship involved in a service transaction: customer-buyers and customer-users. Layers of customers emerge in situations in which the person selecting the service and paying all or part of the cost is buying it for someone else.

This dual relationship presents the service provider with a dilemma, because the customer-buyer and the customer-user can have conflicting needs and expectations. As an example, consider employee benefit plans. Employers, who pay the greater portion of the cost of employee health care and pension funding, are customer-buyers. Their employees, the beneficiaries of those plans, are customer-users. While both groups want the best possible benefits for the price they pay, their definition of what is the "best" may be in direct conflict. This places the service provider in the challenging position of serving two masters.

Dual-customer relationships do not confine themselves to customers outside a service organization. However, in dealing with internal customers, service providers find it difficult to identify the types of customer relationships in which they're involved. Let's eavesdrop now on one of our favorite internal services, the personnel division.

ᨳ ᨳ

"We Answer to a Higher Authority"

"I think that we should begin the meeting now."

John Allen had recently become the senior officer in charge of customer service and performance improvement. After positions of increasing responsibility in many operating areas, John was enthusiastic about his new role. He welcomed the opportunity to bring in fresh ideas and to help the company achieve a better competitive posture.

John continued, "If we want to improve customer service and improve our performance, we have to identify our starting point. So, we'd like to introduce an organizational self-assessment process. During this process, managers identify their major products and services, then name the customers for each and the performance standards they use.

"We've already helped several line operating units successfully go through this process. Now, we'd like to extend it to our staff areas. So, I thought a good place to start was right here with the personnel division—after all, you folks are on the leading edge of human resources thinking. I couldn't think of a better group of people to assist me in the experiment."

Good move, thought Kate. John is trying to win them over with the judicious use of flattery. John had hired Kate Nova to facilitate the implementation process.

John glanced around the table. Sam Nelson, director of personnel, had not shown up for the meeting, nor did he send word that he would be late. "Though I realize that Sam is not here, I think we should begin. We're already twenty minutes late."

"If he's not here, then it must be a crisis," interrupted Lois McMann, the manager of corporate training programs. John chose to ignore the remark and continued with his introduction.

"I'd like to turn the meeting over to Kate Nova. Kate's been facilitating the other groups and is very familiar with the process—in fact, she taught it to me. She's going to conduct the session this morning and begin with a description of the process and its purpose. Kate, I leave the group in your able hands."

John then went to sit at the back of the room, leaving Kate to begin. She had delivered the overview so many times that she could afford the luxury of diverting some of her attention to the people sitting around the conference table. They represented all the major areas of the personnel division: employee benefits, compensation, payroll, compliance review, training, recruiting, and staffing.

Just as Kate finished her talk, Sam Nelson entered the room. "Sorry I'm late—I was with the executive committee," he said to no one in particular. As if on cue, everyone at the conference table turned their heads to follow his movements. He sat next to John at the back of the room, looked up at Kate and added, "Just continue. Don't let me interrupt."

You already have, thought Kate, but she said, "No problem. I was just about to summarize the key points of the process." Kate completed the overview with as much grace as she could. Then, moving a flip chart easel closer to where she was standing, she said, "I think we can begin the process now. Who wants to go first and identify major products and services?"

Kate watched as all eyes turned to Sam. He nodded slightly in the direction of Pat Chan, manager of compensation. Pat turned to Kate, smiled, and said, "I'll start. I enjoy experiments."

Sure you do, thought Kate. "Thank you, Pat. Let's begin." The first part of the process went well enough, with each area identifying its major products and services. The only disagreement was on the definition of the word "major." Kate reminded them that they didn't have to identify everything they did, and the process flowed smoothly after that.

"I think we've done so well on this part, we should go right on to the next step. Let's begin to identify the customers for each of these products and services. Pat, since we started with you, why don't we go back now and name your customers. For example, who are the customers for the job evaluation product?"

"That's easy," replied Pat. "The executive committee." The executive committee's membership included all the major decision makers—in other words, top management.

15

"OK, the executive committee is one customer. Are there any other customers for this product?"

"I can't think of any others," replied Pat.

"They are the *only customers*?" Kate's voice rose in question.

Trying to help, John interjected from his place at the back of the room, "Well, when I was a manager, I remember doing a lot of work preparing job descriptions, reviewing current ones, and going to meetings. The areas I worked in considered job evaluations very important because they had a major impact on staffing and recruiting."

Silence met his remarks. Finally, Sam Nelson spoke. "I agree with my people. The executive committee is the only customer for the compensation area's products. They are the ultimate decision makers on compensation policy."

Kate's heart sank. She needed to talk to John. "We've been working for close to two hours. Why don't we take a break. We've done so well up to now, we're ahead of schedule. Let's reconvene in 10 minutes."

During the break, Kate and John agreed they needed to move into an area where it was clear there were customers other than the executive committee. John recommended the training area. "After all," John reasoned, "they should be able to recognize that there are many different customer groups for each of their training courses."

As the meeting resumed, Kate addressed Lois. "Lois, let's continue with you. Who are your customers?"

"That's easy. The executive committee."

Kate was undaunted. "But what about the employees who attend your skills courses? Aren't they your customers, too?"

"I don't think so. They don't pay for the courses they attend, they don't even decide whether they'll attend or not. They just show up."

Kate knew she was quickly stepping out of her role as a neutral facilitator, but she couldn't help herself. She insisted, "Surely, their managers can be considered your customers."

"All they do is schedule people to come."

Kate pressed further. "But you answer to the managers in some way—" Lois broke in before Kate had a chance to finish. "No, the design and delivery of the courses are left to training professionals who know best—that's why we pay them. The managers don't have anything to do with it. No, we don't answer to them. We answer to a *higher* authority," she said with finality.

❧ ❧

Kate's experience points to a key misunderstanding that internal service staffs commonly share: the myth that the boss is *the only* customer. Typically, bosses set standards, evaluate work outputs, and control rewards and recognition. However, for the voice of the customer to be heard, the boss has to be primarily seen as a surrogate for the true customer.

This point of view reflects the changing role of the boss. In a simpler, less competitive, less market-driven world, bosses planned, controlled, directed, and organized. Work was broken down into simple, repetitive steps, requiring little initiative or original thought from employees. The word of the boss was law.

But the world in which we compete today is more complex, less direct. Technology has transformed the way we work. Employees have changed in character and expectations. Getting the work out requires a different approach. While bosses still plan, control, direct, and organize, they do so as facilitators, catalysts, and coaches. They are sensitive to the needs of their customers and make sure their subordinates are sensitive as well. They strive to have performance standards reflect customer expectations.

The role of the boss in this brave new world is not to be a customer but to be leader of a team that is serving customers. It means that the boss provides his or her team with the help it needs to meet, or even exceed, its customers' expectations. He or she is expected to handle inside problems that block front-line employees from providing service to customers. This demands the ability to coordinate the actions of people who are peers or to influence those who do not directly report to him or her.

For some service providers, this notion is a revolutionary one. It challenges the time-worn assumption that the boss is the only one who knows what is best for the customer. It modifies this assumption, to include those in the organization who work most closely with the customer and strive to understand and meet customer expectations. We'll be considering this point again in our next chapter, when we take a detailed look at the customer-provider work loop.

Question 3: How do we measure the effectiveness of the service we provide? It doesn't matter if you're providing service or manufacturing widgets. If you can't measure what you're doing, you can't manage it. If you don't know what level of service you're providing now, you won't know if you're improving, staying even, or falling behind. So, an important part of the self-assessment process is identifying the effectiveness measures. You can look at these measures as the standards that tell you whether the delivery of a service is under control.

What kinds of effectiveness measures do we use with services? Many of them fall into the same major groups that we'd use when we're manufacturing widgets:

Accuracy: Are we providing the service that we said we'd provide? For example, is the right topping on the pizza? Is the check payment for the correct amount? Did the passenger's bags get on the same flight as the passenger?

Time Service: Are we measuring service against a specified period of time? Do we answer the phone within three rings? Do the restaurant patrons receive the menu as soon as they're seated? Does the computer respond to users within a specific number of seconds?

Productivity: How efficient are we at providing service? For example, how many calls per hour does the 800 operator answer? How many of those calls can be answered completely or must be referred to someone else?

Identifying and tracking these types of measures for each of the services you provide is a starting point. However, while accuracy, time service, and productivity may be enough for manufacturing widgets, they don't provide a complete measure of your effectiveness as a service provider. Why not? Only when you move to the final part of the self-assessment process is the answer possible.

Question 4: How do our customers measure the effectiveness of the service we provide? Before congratulating ourselves that we have a good understanding of the

17

services we offer and the customers we serve and an effective measurement system that oversees the delivery of our services, we still face engaging in an even more rigorous stage of our assessment process. It is also a very meaningful part, because it ties together the results of all we have done thus far.

This stage calls for input from our customers. In a customer-driven organization, these individuals or groups represent the motivating force behind every action we take.

The service provider who engages in customer dialogue for the very first time may be in for a shock. All too often, he or she will find that an incomplete, inaccurate measurement system is in place.

Measurement system failures fall into three major categories: (1) measuring from the customers' perspective, (2) measuring all aspects that are important to customers, and (3) measuring nonroutine activities.

Let's look at each of these categories in detail.

Measuring From the Customers' Perspective

A common trap for service providers is failing to set standards that reflect customer expectations. Fortunately, it's also the easiest trap to spot.

For example, service organizations often set the clock by which they measure their responsiveness to customers from the time they receive a customer request. While this may be fine for face-to-face or telephone transactions, we often find that it is *not* appropriate for processing mail-in transactions. This is because customers often "set their clock" when they drop their request in the mail.

Different perceptions of responsiveness become a crucial concern for service providers who have a third party intervening between themselves and their customers; for example, those who use outside vendors to provide some part of their service or to verify a customer's status.

This type of measurement failure can also be the result of using "industry" or "professional" standards. To the extent that industry groups actively seek and use input from customers in developing their benchmarks, industry standards are fine. However, industry standards often become an excuse for telling customers "We know what's best for you."

While customers may be willing (or forced) to accept these standards for a time, they do not live in stagnant world. Their expectations are dynamic, influenced by their experience with services that, at first glance, seem dissimilar. For example, a customer who calls in an order to a catalogue firm and immediately learns that the item is out of stock will expect to be able to receive the same type of information from other service providers. This customer is unlikely to accept the excuses of a retail store when a promised delivery date isn't met because back order status "wasn't known" when the sale was made.

Measuring All Aspects That Are Important to Customers

Service providers see a service in terms of accuracy, turnaround time, and productivity. These aspects are important but incomplete. They provide only an inward view of the service. To this view, providers need to add knowledge about the service *from the customer's viewpoint.* They need to discover a way of seeing the service *from the outside.*

In any service interaction, customers judge the experience along two dimensions: *what they receive,* which is the end result or outcome, and *how they are treated,* or the process they have undergone. For many services, this process can be just as important as the outcome.

Let's see how the process of a service can color a customer's evaluation.

ﺫﺍ ﺫﺍ

"Get Me to the Plane on Time"

Kate Nova slid into an aisle seat of the lecture hall and looked around for a friendly face. As usual, the Association of Productivity and Quality Professionals (APQP) had crammed the agenda. Members of the other workshops are coming in late as well, she thought. I'm glad my group wasn't the only one with too much to handle.

"Kate, how nice to see you again." George Watson sat down beside her. "What have you been up to since our last meeting? Still trying to convince the engineers that there's more to quality than calipers?"

George was the quality assurance manager for a major petrochemical firm. He remembered Kate's impassioned plea for developing unique service measures at the last regional meeting.

"Now, George, you know I'm not against the traditional approach. I've just been trying to get you quality control guys to see that there's more to service than mere end results."

"The only end result I'm worried about today is our trip to the airport. No one's been able to convince the association to move the regional meetings to a hotel near the airport. Headquarters is the most inconvenient place to get to."

Kate agreed. She remembered the last meeting. George and she became companions in misery as they shared a nightmare ride to National Airport with another member. All three had missed their flights.

"I used the Metro and a taxi from the local station to get here today, George. It took an hour and half, but at least I didn't have to worry about traffic jams, construction detours, and getting lost. And," she emphasized, "the association has a minivan service to get us back to the Metro station. I'm planning to use the three o'clock van ride—that'll allow plenty of time to make the five o'clock shuttle to New York."

"With that guarantee, I'll join you," said George. "I'm on the four fifty-nine flight to Boston. The trip to National will give me plenty of time to pick your brain."

Kate raised her eyebrows and waited for George to continue.

"If I hadn't run into you today, Kate, I would have phoned you. My chief financial officer has been after me to begin a quality improvement program in his operations. Frankly, I'm having trouble getting started. My usual approach is going over like a lead balloon. I know you've done a lot of consulting in financial service companies. Maybe we can arrange a visit and have you give us a few pointers? For your usual fee, of course," George hastened to add.

Kate smiled. "Why, George, don't tell me I'm making a convert of you!"

"We'll discuss my conversion on the trip to the airport," said George, as they both settled back for the reports of the various workshop groups.

At 2:55, George and Kate stood outside the main entrance. By 3:00, George was beginning to fidget.

"If I miss the four fifty-nine to Boston, there's no second chance—it's the last direct flight of the day."

"George, calm yourself. A van has just turned into the driveway. It must be ours."

As the van got closer, they caught a fleeting glimpse of the lettering on the side—"APQP Metro Shuttle"—fleeting, because the van sped past them and screeched to a halt about 50 yards away.

Uh-oh, Kate thought, this isn't starting off too well. Maybe we should have called a cab.

She looked at George, who shrugged his shoulders. Together, they headed toward the van at a brisk pace.

The driver sat stolidly behind the wheel. After a brief pause to see if he would acknowledge their presence, George opened the door and was met by a blast of heavy metal music that rocked him on his heels. He boosted Kate into the back seat and then followed her.

The driver kept his eyes glued ahead, totally ignoring his passengers.

George looked at his watch—3:05. "Excuse me," he began, "isn't this the three o'clock shuttle?"

At that interruption, the driver turned, thrust a clipboard into the back seat, and stalked out of the van toward the lobby of the headquarters building.

"I guess we have to sign this," said Kate, looking at lines that called for her name, the purpose of her trip, and the name of the APQP employee visited. Looking at George's angry face, she added, "I'll take care of the paperwork for both of us."

Ten minutes later, a red-faced George stormed into the lobby. Kate watched his return a minute later with the driver in tow. George certainly looks as if he hasn't won any friends or influenced people, she thought.

Proving her opinion was right, the driver turned the volume of the radio up even higher and began to move the van forward at a snail's pace. Before George could open his mouth to protest, Kate gave him a not-so-gentle kick.

Fifteen minutes later, they were on the Metro, speeding toward the airport. George began fuming almost immediately.

"We'd still be at headquarters if I hadn't gone in after that guy. You know what he was doing? Sitting around shooting the breeze with the receptionist at the sign-in desk. You can bet I'll fire off a detailed complaint letter when I get to Boston tonight—if I get to Boston. Quality and productivity professionals, my foot. If that's the type of performance they tolerate from their employees, it's no wonder our country's quality is going down the drain!"

Kate tapped George on the shoulder.

"George, George, calm yourself. We should make the airport with twenty minutes to spare. But, in the meantime, can I talk to you about service quality again—and the importance of measuring how you treat customers as well as the end results?"

ﻉﻝ ﻉﻝ

Obviously, Kate and George would have been even more upset with the service received if they had missed their planes. As the story illustrates, even when all the outcome measures are met, you can still have customer dissatisfaction due to lack of process measures.

Measuring the Nonroutine

Nonroutine transactions provide the final type of failure in service measurement systems. These are the types of service encounters that strike fear in the heart of service providers, for they mean something has gone awry with the normal service delivery system.

Nonroutine transactions can also occur through no fault of the service organization. These are the situations in which a customer asks the company to make an exception or to "bend the rules a little." For example, under normal circumstances, waiting for five days may be acceptable service for a customer surrendering an insurance policy. However, if the customer's house closing has been moved up to the next day, the customer may ask the provider for 24-hour service.

Why is it important to look at nonroutine transactions? After all, they probably represent a minor portion of the total business. Unfortunately, nonroutine transactions have a very great impact on the customer in proportion to their numbers.

If handled well, they offer service providers an opportunity to make a lasting, positive impression, because customers enter such situations with extremely low expectations. They expect to be mistreated or ignored. In short, they expect the worst.

A service organization also needs to know the type of nonroutine service requests its employees are experiencing to plan improvements. Information on those problems that result from system breakdowns should feed into quality improvement projects. More importantly, however, is using the information to train service employees and to set forth clear guidelines for empowerment. This is a subject that we'll be covering in greater detail in a later chapter.

SUMMARY

The first steps on a journey are sometimes the hardest. On the journey toward a customer-centered organization, the first step begins the process of organizational self-assessment. The four questions posed—involving services, customers, existing measurements, and measurements from the customer's perspective—are deceptively simple; developing the answers is the hard part.

In Chapter 2, we'll look at another exercise that can help in the organizational self-assessment process.

QUESTIONS FOR DISCUSSION

"What Do You Want From Me—I Only Work Here"

1. Is Francine's supervisor sensitive to customers? Why or why not?

2. What does Francine see as the major service offered by HPS? How would she describe her own service? How would you define it?

3. What could HPS do as a company to become more sensitive to its customers?

"We Answer to a Higher Authority"

1. How would you describe the management style practiced in the personnel division?

2. What should be the proper role of the executive committee with respect to the services offered by the personnel division?

3. Consider two of the major services offered by the personnel division—employee benefits and training. Based on your experience as a user, to whom are the managers of the division responsible?

"Get Me to the Plane on Time"

1. What kind of "outcome" standards could Kate and George apply to measure the quality of the service they received during their ride to the airport?

2. What kind of "process" standards could Kate and George apply?

3. In your opinion, does the van company offer a quality service? Give reasons to support your viewpoint.

DEVELOPING A CUSTOMER ORIENTATION: THE CUSTOMER- PROVIDER WORK FLOW

In Chapter 1, we discussed a step-by-step approach to assessing an organization's sensitivity to its customers. In and of itself, the process is a meaningful one, providing insight into organizational values and behaviors.

For example, a benefits unit that does not view employees as customers is unlikely to respond to employees in a caring, courteous manner. Lacking customer focus, such an organization is likely to be driven by policies and procedures ("the rules"), rather than a desire to serve the customer. The unit loses sight of the end customer and all intermediate ones. The overriding goal becomes adherence to policies and procedures.

In such an atmosphere, organizational thinking becomes rigid. Responses to customers become inflexible. Creativity and innovation are discouraged by an attitude that responds to new ideas with "We've always done it this way." Moreover, this type of culture is strengthened by reward systems that reinforce—explicitly or implicitly—behaviors that conform to rules and regulations, rather than reflect the interests of the customer.

The impact of these values, beliefs, and attitudes goes beyond customer dissatisfaction. Organizations afflicted with the symptoms described above suffer from

23

low productivity and high costs, caused by excessive processing steps and checking built into their transaction processing.

How can you tell if your organization has fallen into this abyss? A helpful first step is to examine customer-provider work flows.

THE CUSTOMER-PROVIDER WORK FLOW

This exercise begins where the organizational self-assessment ended. It uses the insight gained from that process as a starting point. The customer-provider work flow examines in-depth the path a service follows from the time its ultimate user requests it to the point at which he or she receives it.

Along the way, all of the interrelationships of internal customer-providers are also examined. The examination probes the following: (1) Who is involved? (2) What exactly do they do to add value to the service transaction? and (3) What standards do they follow?

The result is an understanding of the process—warts and all—by which we currently provide service to our customers.

Before we begin, it's helpful to clarify some key terms used throughout this process. We've already described customers as those who require an output or service to perform their work activities or because they are the ultimate users. Providers, on the other hand, are those who supply the output to customers. For example, the instructor who delivers a course is a provider; the students in his or her class are customers. The bellhop who carries the guest's bags is a provider; the hotel guest is the customer.

The term *customer-provider* acknowledges one of the unique characteristics of services, namely, that customers are an integral part of the service transaction. The input provided by customers affects the service provider's ability to supply satisfactory service. Indeed, the whole service transaction is colored by the intimate participation of the customer.

In internal customer relationships, this is easy to see. For example, consider the interactions between a salesperson and the order processing clerk who will complete the paperwork needed to close the sale. In this instance, the salesperson relates to the clerk as a provider, because he or she provides the input that the clerk needs to process the order. However, once processing is complete and the clerk returns the completed order form to sales, roles are reversed. Now, the clerk is providing an output that the customer—in this case the salesperson—needs to complete his or her work.

Even the ultimate user, who is usually an external customer, plays the role of "provider" in initiating a transaction. The provider role played by the external customer is often overlooked or poorly understood. Let's look at a familiar situation that will better illustrate the customer's role as a provider at the start of customer-provider work flow.

Think back to the last time you had dinner at a restaurant. From the minute you entered the restaurant, you provided the restaurant's employees with important information, so that they, in turn, could provide the type of service you wanted and expected. For example, you told the waiter how well done you wanted your meat and what vegetable you preferred.

24

In this instance, you were the provider of information and the waiter who received it was *your* customer. The waiter, in turn, "processed" the information, writing "medium rare" and your vegetable choices on the order slip.

At this point, the waiter became a provider, as the completed order went to the next "customers" in the work flow, the folks in the kitchen who would prepare and assemble your order. They, in turn, became providers, passing their processed input along to their own customers—the next step in the work flow.

This pattern of customer-provider relationships continued through the work flow until you, as customer, ultimately received your dinner from your direct provider, the waiter.

Figure 2.1 shows the customer-provider work flow for this restaurant operation. Everyone involved in the service interaction plays a dual role: provider and customer. It also illustrates other important characteristics of the service relationship.

First, the pattern of relationships can be seen as a closed chain or loop. For the service transaction to take place, there must be closure. The ultimate user in our example will not receive the order if a link in the chain breaks or is interrupted.

Second, each customer-provider within the flow must receive high-quality service if the ultimate customer—the person who initiated the transaction—is to receive high-quality service. If the waiter writes down the wrong vegetable for your order, the chef cannot provide what you wanted. Thus, the service you receive reflects the total value added by every link in the chain.

Third, the same link in the chain may be involved at more than one stage in the work flow. Getting back to our restaurant, the waiter was involved both at the beginning, in taking your order, and near the end, in delivering the food. For more complex services, especially those that involve white-collar "paper factories," a transaction may return many times to the same service person. At each pass, the service person plays several different customer-provider roles.

Finally, the customer-provider relationship does not exist in isolation, even though, for simplicity, it is shown as a closed loop or chain. Instead, it probably connects to many other chains, which belong to earlier customer transactions. This interdependence can have an important impact on later interactions. In our restaurant illustration, a prior customer-provider interaction would have taken place when you called to make a reservation. You acted as a provider of information, indicating the date and time, the number of people in your party, and which section of the restaurant you preferred—smoking or nonsmoking.

At that point, the person taking your reservation was a customer, who then processed your information by making the appropriate entries in the reservation book.

On the day of your dinner party, the person greeting you at the door was a customer of the person who originally entered your reservation. Obviously, a mistake made in processing the reservation would have a negative impact on the service you receive and might cause you to cancel.

Most of us identify with the experience of a restaurant customer—we easily identify the customer-provider work flow and the impact these interrelationships have on us as customers. As managers or individual service providers, however, we sometimes lose sight of this chain, especially in a white-collar service. We're biased by

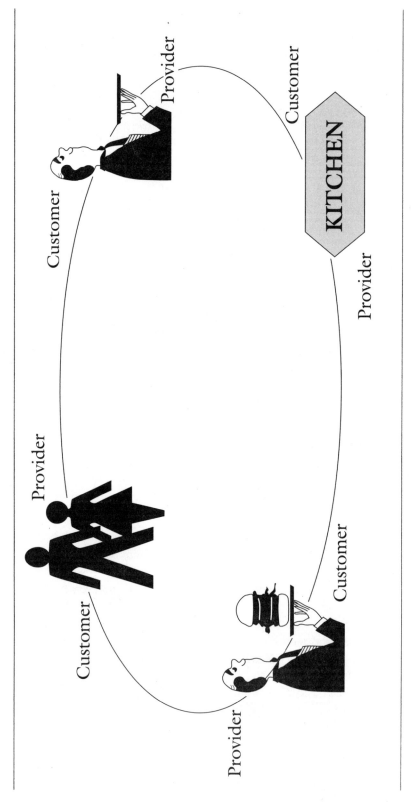

Figure 2.1: The Customer–Provider Work Flow

our close involvement in the work process and by our isolation from ultimate users. We become so familiar with the assembly-line nature of the work flow, we fail to recognize the individual customer-provider interactions that take place all along the way.

Often, it's helpful to stand back and observe the whole landscape of interactions. Let's visit with a typical back office manager and see how well he and his people understand customer-provider work flows.

ಿ ಿ

"If Only We Didn't Have Such Stupid Customers!"

Bill Smith hung up the phone with a sigh—another upsetting call from an irate customer. Lately, it seemed that every other call that reached his office was a customer complaint about how long it took to get anything done.

He glanced up at the graphs posted on his bulletin board; they showed a steady series of standards met by each of the units reporting to him. Yes, each one was meeting its time service requirements, but obviously it wasn't enough.

Well, this place is just a glorified paper factory, he thought, so I might as well begin at the beginning. He left his office and headed toward Gracie Aldon, the team leader of his data assembly unit.

As always, the members of Gracie's team were bustling around. Gracie was a long-time employee who had risen through the ranks to her supervisory position. She was known for running a tight ship—no goofing off under her command.

"Gracie, can we get together for a few minutes? I'd like to discuss the customer complaints we're getting about our turnaround time."

Gracie frowned at Bill as if he were a mailboy she'd caught reading *Sports Illustrated* while the mail went undelivered.

"Who's been complaining about my team?" she challenged him.

Bill decided a large measure of diplomacy was needed.

"Gracie, it's not your team in particular, it's the whole process that has to be reviewed. Your folks are such key players, I thought your team might be the best place to start. After all, you're the ones who are first to get customer requests when they roll in the door."

He continued, "I'm not worried about the majority of cases that your team can pass on to the customer services unit. It's the other kind that get put on call-up that I'd think we should look at. The records show that's about thirty percent of what comes in. Thirty percent of twenty-five hundred a month adds up to a lot of cases."

"Leave it to you to highlight the problems, rather than talk about the good things we do," Gracie replied. "Anyway, I told you, we don't call them call-ups anymore—we call them RIPs now."

"Oh, that's right, I forgot. Let's see, that's 'requests-in-progress,' right? But Gracie, changing its name doesn't solve the problem. We've known each other too long to let a name get in the way of understanding. We might be able to cut down the complaints if we knew the reasons why so many requests are put on call-up."

"I just don't know why you keep bringing this up. My team is already meeting the time service goals you set up a few years ago. Besides, everyone around here

27

knows why we can't shorten the turnaround time—it's the customer's fault! We have to keep writing back to them because they don't know how to fill out the forms. Simple things like including the subdivision codes or even just signing the request. You'd think after all this time, they'd have learned."

She paused before continuing, lowering her voice. "Of course, if I had a phone unit on my team, we could get some of the missing information more quickly."

At this point, Bill was privately agreeing with those who believed in reincarnation. He knew without doubt that he had lived through this conversation before. Two weeks ago when he had raised this issue, Gracie suggested forming a phone team on her unit. She pointed out that phoning for missing information is quicker and ends the need to write back in every case. Bill rejected the idea then, as he did now.

"Look, I just don't have the kind of budget that allows for more staff or the extra telephone charges." He omitted telling her that he was also reluctant to ask upper management to fund an unproved idea.

As Bill stood up to leave, she added, "We're doing the best we can with the information we get. Besides, I don't know about the other units, but as I said before, *my* unit is already meeting its goal." She hadn't meant to sound so defensive, but knew she hadn't succeeded.

&. &.

Although it may appear illogical, Gracie was right when she pointed to the external customers as the source of her problems. Because they were initiating service requests, the customers were actually playing the role of "providers." They were responsible for supplying the information that Gracie's unit needed to start processing the transaction.

The lack of information prevented Gracie's team members from doing their jobs. They couldn't add value to the transaction without information that they relied on their customers to provide. In fact, for 30 percent of their work load, they were incurring extra costs: extra time spent writing the customer for additional information, time spent filing and refiling, and extra postage charges!

How do we help external customers become informed "providers" as they initiate a service request? Gracie offered one obvious, although costly, answer. She proposed a telephone unit that would call customers for the missing information. However, that is still fixing something *after* we discover a problem. The more effective solution is to *prevent* the problem from happening in the first place—*before* the request ever reaches us.

By setting up processes and procedures that educate the customer, this can be accomplished. Simply shortening and redesigning forms makes it easier for customers to complete them. This alone can prevent many errors. Another way is to include reminders on the flaps of the envelopes customer use to remit payments. The key is to make the customer's interaction with us as painless as possible and to carry this goal throughout every step in the work chain.

THE ROLE OF INTERNAL SUPPORT STAFFS

The same type of problem that Gracie had with her "external providers"—inadequate or inaccurate information—can take place along every link in the chain, multiplying the costs involved and adding extra time to the processing cycle. A provider playing a role early in the work flow may contribute to processing problems. He or she may be called on to supply information needed by a customer who is much further along in the process or who is part of an interlocking work flow. If the provider cannot see a reason for supplying the information or understand how it is used, the information may be omitted or be carelessly provided. The result? Serious problems for other internal customer-providers.

For example, let's consider a payroll clerk's job. The primary responsibility is to enter the proper data so that employees receive the correct amount of pay in a timely manner. But the clerk is also called on to enter job code information. The code has nothing directly to do with the calculation of pay or taxes. It is needed by a programming unit that produces monthly hiring reports from the payroll files.

Of course, the payroll clerks *and* their supervisor are primarily concerned with the data needed for calculations. The standard by which their performance is judged is the accuracy and timeliness of the paychecks they produce for their ultimate customers—the employees.

It is easy for the payroll unit to ignore the needs of the programming unit. After all, the programmers are not a direct part of the payroll transaction customer-provider work flow. They belong to an interlocking chain. If the payroll clerks do not recognize this relationship, they may not be as careful with job code data entries as they are with the payroll data. As a result, the programmers may not receive accurate information for their hiring reports.

Customer-provider work flows help internal customer-providers understand their roles in the work process. Units and individuals can begin to understand the "big picture." They also begin to understand their dependence by recognizing their relationship to others in the work process.

For internal support units to survive, they *must* understand the role played in providing service to *all* customers. This means not only looking at the next customers in the work flow but also weighing the needs of other units in the chain, as well as the expectations of ultimate users outside their organization.

Bill Smith's next encounter shows the dangers involved when an internal staff fails to develop this type of understanding.

&. &.

"Don't Bother Me—I'm Doing My Job"

Bill Smith picked up the phone and punched in a familiar number. At the other end, Emily, supervisor of the calculations team, glared at the ringing telephone. She had just cradled the receiver a second ago, and now it was ringing again. Is everyone else's life like this, she thought? For a moment, she toyed with the idea of disconnecting the phone; instead, she picked up the receiver.

"Calculations, Emily speaking."

"Hi, Bill Smith here. How are you doing, Emily?"

"Fine until I heard your voice, Bill."

He ignored the irony in her voice. "Emily, where are the calculations on those priority requests Tom White sent to you four days ago?"

Tom was Bill's fellow manager who had the accounting and data processing support functions as two of his major responsibilities.

"Never heard of them," she lied.

"Come on now, Emily, you know Tom needs your team's numbers so his accountants can close out the monthly reports."

"Tom only gave them to us late last Friday and today is Wednesday. I've got twelve people on overtime already and the backlog isn't getting any lower. What more do you want?" Emily retorted.

Bill could feel himself tensing up. "Look, we both know that not all your clerks actually do requested calculations. You've got four checkers who redo everything the calculators do. We could probably shave two or three days off the time it takes to process a request if your team did less checking and more work."

"That's right, Bill, criticize us for producing at one hundred percent accuracy. Expect a promotion to vice president soon—you're beginning to sound just like those guys in corporate headquarters! By the way, when did you start working for Tom White? That little sneak…"

He interrupted her. "Look, Emily, see what you can do to get those requests completed today—even if you have to do them yourself." He hung up before she could argue with him.

Bill resented the way Emily treated him, but, to be fair, he realized the situation wasn't entirely her fault. She and her team were pressured from all sides.

On the other hand, Bill knew that Tom felt victimized because his team's work depended so heavily on the calculations team, whose priorities differed from his own. Lately, Tom had been hinting about reorganizing to form his own team of calculations clerks.

The only reason he hasn't brought it up at the monthly managers' meeting, thought Bill, is the cost. Tom couldn't justify the expense involved in adding more staff to his unit, and he wasn't anxious to do battle with Emily over transferring some of her clerks to his organization.

Even if I were to agree with Tom, Bill thought, the solution is more trouble than it's worth. Reorganizations are so messy—we'd have to get layers of management approvals, and then we'd have to grapple with the personnel division as well. He hated the thought of all that paperwork—and even if they did all that, what proof was there that the change would improve anything?

Better leave well enough alone.

&. &.

USING CUSTOMER-PROVIDER WORK FLOWS
TO IMPROVE ORGANIZATIONAL EFFECTIVENESS

The situation that Bill, Emily, and Tom faced is a familiar one. Employees in organizations that fail to understand customer-provider work flows don't see the "big picture." They don't see all the other people who depend on a particular work activity, and so it becomes difficult to set priorities. The norm becomes reacting to problems, rather than preventing them.

Developing customer-provider work flows is not merely a tool to better understand customer needs and expectations. It also serves as an important technique for pinpointing the parts of the work flow that are ripe for improvement. The first step is to take each of the major products and services that were identified in the organizational self-assessment described in Chapter 1 and begin to develop a customer-provider work flow.

For each service, look at the customer(s) involved. (1) How does this customer begin the service request? (2) What person or unit does he or she come into contact with first? and (3) What must he or she provide during this first step to enable the next step in the process to take place, the next link in the chain to be forged?

Then, put yourself in the place of the first "customer" in the organization who receives the information. In this role, think about the following: (1) What must you do to complete this part of the process? (2) What must you provide to your "customer"—the next person in the work chain? (3) How often do you receive incomplete or inaccurate information from your "provider"—the customer outside your unit who initiated the request? and (4) When you receive poor information, what actions do you take?

Continue this process all along the chain, until you reach its close, where the initiator—the person who originally requested the service and provided the initial information—now receives a response. At this point, the provider and customer are the same.

Along the service path, you meet many other customer-providers. At each step, estimate the time needed to complete each step if the "customer" handling it has been provided with all the information needed to do his or her job. Then, estimate how much time it takes to complete the step when all the necessary information has not been provided. Identify the extra resources consumed, such as paper, postage, phone calls, or computer time. At the end, add up the time and costs involved for "clean" transactions, and do the same for those that aren't clean. This comparison provides a starting point for operational improvement.

Examining the work flow also provides an opportunity to review the details of the process. Look critically at each activity in the work chain. Ask yourself the following questions: (1) How many links in the chain represent checking and rechecking? and (2) Why are they being done?

Excessive checking may be the result of a knee-jerk reaction to a type of error that was common in the distant past. Or it may be a symptom of serious difficulty with the input received from providers along the way, suggesting a need for "provider education."

After developing a customer-provider work flow, it is important to revisit it regularly. Service relationship chains do not exist in isolation. They directly or indirectly affect service interactions that occur elsewhere in the organization.

A key challenge, then, is to identify interlocking chains along your primary customer-provider work flow. For internal staffs, this may be the only way they can establish their relationship to customers outside the organization. This step adds meaning to their work and heightens individual employee motivation.

Many organizations contain parallel units that provide the same service or do the same work. Go through this exercise independently with two or more of the units. This often results in major surprises, as Bill Smith begins to learn in his next meeting.

ૐ ૐ

"Our Only Aim Is to Serve Our Customers"

Bill couldn't think of anything he disliked more than having another exasperating meeting with Michael, the newly appointed supervisor of the customer service unit.

"What progress have you made on the customer representative training program?" Bill asked. "You've got two new reps and six or seven more with less than six months' experience. Your unit has always been the centerpiece of our service to customers, the one with the most direct contact. We have to get your reps up to speed as soon as possible."

"I've got the benefits quote and certification procedures pretty well documented. Now I think I can start the workbook sections on these pieces."

"I thought those pieces were completed weeks ago," Bill said, trying not to let his voice betray his frustration.

Michael shifted uncomfortably in his chair.

"Documenting procedures in my unit is harder than I expected. All the reps seem to do things their own way—even the senior reps don't agree on basics. Sometimes I think that my three units are each working for a different company. On top of that, I'm not sure that the procedures we use are helpful to customers. We seem to do things that make life easier for us but that make it harder to do things for customers."

"Besides, I've been spending most of my time on the first draft of the letter answering the charges made by that troublemaker in Delaware."

Bill's effort at composure slipped.

"Stop whining, Michael, I can't stand it—and don't call her a troublemaker! She's a customer and upset at the treatment she's received. I thought you were more sensitive to customers. After all, you were a customer rep yourself for three years.

"Anyway, why are you writing the first draft? I thought we agreed that one of the senior reps could do that. Look, I don't care how you do it, but the next time we meet, I expect you to give me those workbook sections in final form."

Bill watched his newest supervisor leave his office. Had he made a mistake making him the supervisor? Everyone in the office agreed that Michael was an excellent customer representative. What did he mean we make life easier for ourselves instead of looking out for the customer—what kind of a remark was that?

The company had an important responsibility to its customers, which is why all those time-consuming procedures were in place. Procedures were procedures, rules were rules. It was all for their protection. Maybe I did make a mistake after all, Bill concluded. How can Michael be a good supervisor when he just doesn't appreciate what the company does for its customers?

❧ ❧

Pity poor Michael, the new supervisor. He was learning the hard way that not all units doing the same work necessarily do it the same way. He was also realizing that not every step his people took added value to the overall process. In fact, he discovered that some procedures made it more difficult for his customers to deal with the company.

How can Michael use this insight to sensitize his manager to customer needs? Creating a set of customer-provider work flows for each of the units is a major first step. For internal support staffs, this exercise can be an eye-opening experience. Frequently, it represents the first time their relationship is traced to the ultimate user—the paying customer outside of the organization.

It's important to realize this is not an exclusive technique reserved for technical experts. Its power comes from including all members of the organization. Not only will they better understand their importance as customer-providers, but they will also more easily identify the bottlenecks in the current process. After all, they face them on a daily basis! In addition, for a supervisor like Michael, overwhelmed by daily problems, the participation of his team members provides a welcome relief.

But participation is often easier said than done. The challenge is *effective* involvement. This means helping *all* members of the organization to contribute in a productive, meaningful way. In Chapter 6, we'll provide some guidelines on how to do this.

SUMMARY

Developing a customer orientation requires more than a knowledge of the expectations of our external customers. It requires a comprehensive understanding of the requirements of all those who are involved in creating and providing the service.

The customer-provider work flow introduced in this chapter is useful in meeting this need and benefits the service organization in several ways:

1. It identifies all steps that must be taken to produce a given service.

2. It highlights the dual role—customer and provider—that is played by each person or unit involved in the process.

3.   It helps internal staffs to see how their activities have a direct impact on external customers.

4.   It provides insight into dependence between internal staffs. This can be the first step to further the understanding and cooperation between units that have been at loggerheads in the past.

5.   It can pinpoint immediate opportunities for improving operations, including excessive checking, redundant activities that add little value to the end product and missing steps that create problems for other units.

In Chapter 3, we'll discuss ways in which analysis and measurement tools can help us create a more effective service organization.

QUESTIONS FOR DISCUSSION

"If Only We Didn't Have Such Stupid Customers!"

1.   What kinds of service measures do you think Bill is using? How did he arrive at them? Would his customers agree with them?

2.   In Gracie's unit, poor information from customers causes extra steps to be taken in 30 percent of the cases. What can be done to reduce or eliminate this percentage?

3.   Think about your own unit. Are there situations in which better information from your customers would simplify your processing? What can you do to accomplish this?

"Don't Bother Me—I'm Doing My Job"

1.   How does Emily judge the quality of work done by her unit? What methods does she use to achieve these standards? What does it cost her to do so?

2.   Who are the customers of the calculations unit? Do they judge the quality of what they receive in the same way that Emily does?

3.   Describe the customer-provider work flows that exist within the calculations unit.

"Our Only Aim Is to Serve Our Customers"

1.   Why is Michael having trouble coping with his new job responsibilities? What can be done to help him?

2.   Who is more "customer sensitive"—Bill or Michael? Why?

3.   The three units reporting to Michael perform the same work, yet each follows a different procedure. What are some of the reasons why this may be happening?

CREATING AN EFFECTIVE ORGANIZATION USING ANALYSIS AND MEASUREMENT TOOLS IN SERVICES

For a service organization, improvements in customer service demand that we explore opportunities to increase our organization's effectiveness. In Chapters 1 and 2, we describe the necessity of keeping everyone's eyes and ears open to the customer. Serving the customer is our reason for being and our motivation to cooperate with each other.

However, the customer can never know our organization as well as we do. In fact, most of what happens there will remain a mystery to customers. Recall the example of the restaurant described during our discussion of the customer-provider work flow. The restaurant's customers deal only with the customer contact employees. They don't interact with the kitchen staff (except in rare cases to meet the chef), nor do they meet with the people employed to wash the pots and pans. They are even less likely to meet the staff who clean the dining area and make sure clean linens are available.

In fact, services have a lot in common with stage productions. An audience (the customers) sees only what is presented on stage. The performers, however, are not acting alone. They are supported by many others who make critical contributions to the production's success. These backstage people (the theater's back room operations) outnumber the cast. Without them the production could not become reality. Yet the value they add is not directly apparent to the audience—nor should it be. Part of any production's success depends on the subtle artistry of the support people to create "magic," that is, to make their contribution invisible to the audience and everything flow seamlessly.

Furthermore, the general audience evaluates the whole production in terms of the stage experience. Just like audiences at the theater, customers of a service are interested in the results of what happens inside the organization and how they are treated when they deal with customer contact employees. They simply (one is tempted to say selfishly) aren't interested in how difficult it is to produce.

For these reasons, we believe that customer satisfaction is the best measure of organizational effectiveness in the long term. This is as true for nonprofit organizations and government agencies as for profit-making enterprises. We are aware that there are other measures of organizational health: cost-effectiveness, profitability, having money for research and development, gaining and retaining more customers, and so on. These, too, are important overall signals that things are going well or ill. However, it is only by managing for service quality that an organization can survive the rough times of budget cuts and cyclical market changes. Using customer satisfaction as your goal allows you to invent ways to provide quality service at lower cost—that's giving value to the customer!

Customers can tell us what they expect from our products and services. They can even differentiate between what they will live with as opposed to what they desire ideally. What they can't do is help us translate those needs into the most effective internal processes for our organization. It's up to us, as service providers, to remain constantly dissatisfied with our current situation. Seeking fresh approaches helps us to serve customers better by making our internal processes, policies, and procedures as efficient and effective as possible. It keeps our costs down, gives us the opportunity to produce services with more value to our customers, and provides fertile avenues of innovation for employees to explore as they participate in decisions affecting their work. This involvement, in turn, creates a sense of meaningful work for employees. In short, everyone benefits.

Although we have made the case that customer satisfaction is the ultimate measure of organizational effectiveness, we realize that other internally driven measures help us move the organization in that direction. The remainder of this chapter addresses some practical approaches to achieving effectiveness and efficiency in service organizations.

MANUFACTURING TECHNIQUES

Quantitative techniques and measurement tools developed in the manufacturing sector help point the way to identify areas for operational improvement. They provide good controls over work flow. Although originally created to measure products coming off an assembly line, many of them can be adapted to services. This is especially true for services involved in mass transactions, such as banks, credit card services, and retail mail-order companies. Individual experts, managers, and teams benefit from the use of these tools because the results of the analysis allow them to focus on concrete, actionable information.

Some techniques we have found to be useful are the following:

1. *The "80/20 rule" or Pareto analysis,* which highlights the important few problems in contrast to the insignificant many. This approach can be applied to many types of service problems (for example, complaints, errors, and reasons for phone calls) and helps to identify priorities.

2. *Cause-and-effect analyses.* Two of the most familiar are the "fishbone diagram," in which probable root causes are diagrammed to show all cause-and-effect relationships, and the "force field analysis," where problems, their causes, and the forces driving and restraining them are explored.

3. *Trend charts,* which graphically depict such indicators of performance as production, over time.

4. *Sampling* to ensure that the service is meeting customer expectations. This is a cost-effective alternative to inspection. If properly done, it points out variations from predetermined averages or ranges of acceptable output.

We could go on adding to this list for many pages. While most of these techniques were originally designed for use in manufacturing, services can and should explore their use. Education for improving a service organization's ability to create value for customers should include a familiarity with all these basic tools and techniques.

None of these tools requires the use of a computer or someone with an engineering degree to apply them. Any manager, supervisor, or small business owner can successfully use these tools to achieve better outputs. Remember to remain flexible in your application of any specific method. The technique may be the same one a manager in a factory uses, but how you apply it and interpret the results may be unique to your service situation. Keep in mind that the end result is a more effective organization, not the impeccable execution of a technique. All the colorful trend charts in the world do not necessarily lead to more satisfied and loyal customers. Challenges exist at every level to invent new ways to use, change, and adapt analysis tools for your particular organization.

SCRAP AND WASTE IN SERVICE INDUSTRIES

One way to make organizations more effective is to increase their efficiency by eliminating waste and redundancy. The problem with services is that it is sometimes difficult to see the waste or scrap produced by giving the customer something incomplete and doing something over and over to get it right.

These are typical inefficiencies of services and particularly of white-collar organizations. In manufacturing, the scrap heap is tangible: You can see it, weigh it, touch it. If you're not careful, you're likely to trip over it! Services create an "invisible scrap heap," which is no less real for being invisible. It is intangible and may not be an object at all. Thus, you may not be able to weigh or touch it. Yet, waste exists, causing customers many problems. It greatly increases cycle time and decreases the responsiveness of any organization in the customer's eyes. Ask anyone what he or she hates about dealing with a large bureaucracy and among the answers will be, "They take forever to do anything!" The waste caused by not doing things right the first time reduces financial health any way you measure it, whether by runaway costs or defecting customers.

Unfortunately, when defect and error rates are high, the conventional approach of most manufacturing operations is to weed out the bad products and send the good ones to the customer. This is also true for services. It is commonly called "the cost of doing business," as if producing the wrong things were normal business practice. (Unfortunately, this is the case for many services.)

The most expensive way to deal with errors or defects is to inspect the quality of what you send out to the customer. Another way of dealing with errors is to do things over and over until they are eliminated. We call this extra-processing and define it as follows:

Extra-processing occurs when things are not done right the first time. This is true no matter who causes the do-overs, even the customers.

Perhaps the best way to illustrate these points is with another story.

⋅⋅ ⋅⋅

"Only Factories Have Scrap Heaps"

"Kate, it's nice to see you again."

Alice Walsh, the manager of the claims payment division, was waiting for Kate Nova at the entrance to the office. This was Kate's third visit, and the two had an easy working relationship. Kate admired Alice as a good, hands-on manager, whose people both liked and respected her.

For her part, Alice looked to Kate and her experience as a consultant for help in identifying opportunities to improve her operations.

Alice added, "Do you mind waiting a few minutes before we go upstairs? I need to stop by checkwriting to see if their system problem's been ironed out."

"Don't worry," Kate replied. "I can put the time to good use. I need to make an appointment with Loretta in the mail room."

Kate crossed the hall, thinking about her next steps. Alice and her people already had a strong sense of their customers' needs, and most of the staff had ideas on how to improve things. But, Kate asked herself, where should they start? They need to work on something that will give them a sense of accomplishment, that will encourage them to continue.

Now, if this were a factory, I'd have no problem advising them. They'd be tripping over the scrap heap themselves! It's easy to find the pits in a steel mill—I've got to find where the waste is lurking *here*.

"Hi, Loretta," Kate said to the mail room supervisor. "Sorry I couldn't get back to you before this, but I was hoping you'd have time for me on this visit."

"We're looking forward to talking to you, Kate. You know, we do an important job here. If we weren't taking care of the mail that comes into this place, the approvers wouldn't have anything to do!"

Loretta pointed proudly to her team of clerks. As one dumped out large bags of mail, others ran the individual pieces through letter opening machines. The rest of the team quickly reviewed and placed each item into one of three piles.

"I certainly see that you're doing your best to keep the approvers in business," said Kate. "Three different piles—how do your folks know how to sort all that mail?"

"Well, the first is for clean claims."

"Clean claims?"

"Yes, those claims coming in to be processed for the first time. Then, the second pile is mail."

"Mail? I thought it was all mail," said Kate ingenuously.

"Yes, but this is mail with questions about a claim that was already processed or letting us know that something is wrong. A lot of them are answering our letters—you know, where we had to ask for more information or where the doctor forgot to put in the diagnosis. Then, the last pile has returned checks. That happens when the check is sent to the wrong address or returned because we've asked the claimant to return it."

"Why do that?"

"Because it means that we either made a wrong payment or sent the check to them instead of the doctor or hospital."

For a moment, Kate was left speechless. I think I'm having what is called an 'aha' experience, she thought. It occurred to her that other than the pile with the clean claims, she was staring at this office's equivalents of a scrap pile.

"Are you all right, Kate?"

"Of course," she replied and smiled. No wonder Loretta thinks there's something wrong, I'm standing here with my jaw on the floor.

"Look, there's Alice waving to me, telling me she's ready for our meeting. Can we talk some more after lunch?" After Loretta agreed, Kate joined Alice.

On the way upstairs, Kate said, "You have a phone unit in this office, don't you?"

"Sure, the 'complaint unit' as we call it."

"So calls come in from claimants who have questions about their claims or disagree with what they were paid?"

"Not only claimants, but doctors and hospitals, too."

They walked into Alice's office and made themselves comfortable. "Well, Kate, this is your third visit. I know it's early in the study, but do you have any preliminary ideas on what we need to focus on?"

"I think I have a way of measuring the amount of waste, what I call extra-processing, in your office. That would be a big step because it would take this discussion from concepts to concrete examples—something that you and your employees can really work on to improve service while reducing costs."

"I'm all ears, Kate," said Alice as she moved closer.

"I just noticed two piles of mail, one with correspondence and another one for returned checks. These two, added to all those phone calls, are the measure of your extra-processing costs—the price of doing things wrong. It's like a scrap heap in a   41 factory." Kate felt the color rise in her face. What a powerful idea! She waited for Alice's enthusiastic reply.

Instead, Alice looked troubled. "But Kate, in a factory the scrap heap comes about when someone or something in that factory screws up. This isn't a factory. A lot of what is in those piles, including the phone calls—well, it's not our fault! We can't do anything about it. Sometimes, the customers don't give us the right address, or sometimes the customer doesn't sign the form or tells us to send the doctor the check, then changes his or her mind. I tell you Kate, most of it is out of our control!"

This was not the reaction Kate expected. She hurried to salvage the situation. "Alice, it was never my intention to place blame or fault, believe me. But being able to identify the extra-processing is a big step forward. From here, we can begin to identify why the checks are being returned, for example, and address that problem. We now have an opportunity to get at root causes."

"Well, OK, Kate. I'll go along with your idea, but I'm very skeptical about how this is going to be received by the rest of the staff. You're asking them to solve problems caused by others. That's demotivating."

Kate took a deep breath. "I think we should take this slowly and tackle things systematically. Let's focus on doing things better for the customer, rather than who's causing problems, OK?" Kate smiled at Alice, looking for some agreement.

"OK, Kate, I'm willing to listen—if you are, too."

Well, that's a start, thought Kate.

&a &a

As our story illustrates, the use of resources for fixing things up or responding to complaints adds a great deal to the cost of producing a product or service for a customer. Yet, this source of waste is not appreciated (or even acknowledged) by those who work with it every day.

REVISITING THE CUSTOMER-PROVIDER WORK FLOW

Another practical way to identify waste is to revisit the customer-provider work-flow discussed in Chapter 2. Most services consist of numerous steps, which adds to the difficulty in "seeing" the scrap heap. How many steps depends on the nature and complexity of the service. As we discussed before, services frequently have complex "back room" operations. These people produce what is needed to support customer contact employees and often outnumber front-line people. The process can also involve resources, expertise, and other inputs from outside the organization.

According to our definition of extra-processing, it doesn't matter who causes the problem. If something has to be done over, then it adds to the cost of doing business.

In our story, Alice accepted that customers caused some of the problems. She resigned herself to the "normal" way of doing business. Kate, our consultant, realized that the recognition of extra-processing (waste) provided a wonderful opportunity for people inside the company to analyze the situation and invent a better, more efficient way of doing business. The story points to the potential benefits of involving the people who do the work. Their insights can yield impressive results, while helping them feel more in control of and accountable for their work.

The customer-provider work flow examines the path a service follows from the time its ultimate user requests it to the point at which he or she receives it. At that point, we examined each step to help identify the provider and customer at each point in the process. The same work flow can also help us begin to analyze areas of potential extra-processing.

When we look at the work flow this time, however, let's apply a microscope to the work flow itself and get more detailed information about the particular transaction. For example, ask the questions listed Table 3.1:

Table 3.1 Potential Areas of Extra-Processing

1. Does the work pass through the same unit many times? Why?

2. Does the work require customer input at one or more points in the middle of the process?

3. Who supplies input to the process? At what steps?

4. Is information or other input required from outside the primary organization?

5. What role do customer contact employees play in the work flow?

Asking questions like these results in a clearer understanding of the work flow. This alone is a significant breakthrough in an organization, as people involved in the process see the process laid out before them as a whole: connections, break points, and so on.

WHERE TO LOOK FOR EXTRA-PROCESSING

Once you have charted your work flow, you can begin looking for the do-overs. Extra-processing can take place anywhere: before, during, and after producing what goes to the customer. The source of the problem may occur outside or inside the work flow. For example, customers are frequently a source of extra-processing. This is particularly true if communications with customers are restricted, severely limited, or never include educational messages. Some work processes inherently create the need for do-overs, especially those that have built-in controls for security.

Do-overs are also caused by something not even seen in a charted work flow: communications. When something ricochets back and forth between providers, the problem could be the way they are talking to one another. Gaps in internal communications block the coordination needed across the organization to produce a reliable result.

Finally, looking for extra-processing should include those involved in the customer-provider work flow. The people involved at each step are themselves providers and customers of what took place before. They receive and work with inputs and resources from outside the process. There are no better sources for discovering and eliminating the causes of extra-processing.

SUMMARY

We acknowledge that the recipe for achieving service quality described in this book is not glamourous. The front pages of major newspapers and the nightly news broadcasts are filled with stories about the latest mergers, acquisitions, or international agreements. In contrast to these media-focused events, customer service and quality appear mundane, even dull. Stories about creating and implementing processes and programs that focus on the customer and make it easier for more people to participate in decisions never earn banner headlines. Achieving service quality requires hard work, creativity, patience, and a commitment to a long-range vision.

QUESTIONS FOR DISCUSSION

"Only Factories Have Scrap Heaps"

1. What types of extra-processing occur in the claim payment division? For each type, suggest how one or more of the quantitative techniques (e.g., Pareto analysis and cause-and-effect diagramming) can be applied by Alice and her people. Focus on identifying problem priorities, causes, and solutions.

2. How can Alice and her staff begin to identify the "cost" of extra-processing? What cost components must they consider?

3. Do you accept Alice's contention that working on problems that are not her staff's fault will be demotivating? Why or why not?

MEASURING QUALITY FROM THE CUSTOMER'S PERSPECTIVE: HOW CUSTOMERS VIEW SERVICE QUALITY

Because of the unique characteristics of services discussed so far, a service provider's ability to use traditional quality control tools and techniques is limited. This means that providers cannot easily evaluate their service. Unlike a product, a service cannot be measured in isolation against a precise and objective set of standards. Just as beauty is in the eyes of the beholder, service quality is measured through the eyes of the customer.

Organizations that provide services face obstacles unknown to those that produce pure goods. These obstacles limit the extent to which traditional quality tools and techniques can be applied in a service environment. They derive from the very nature of services, the characteristics that set them apart from goods. These, in turn, have an impact on the way customers judge the quality of service.*

*The authors are indebted to the work of Len Berry, Parsu Parasuraman, and Valarie Zeithaml for the ideas and concepts presented in Chapters 4, 5, 7, and 8.

How does a service provider overcome these obstacles? This chapter and Chapter 5 will explore ways that services differ from manufacturers and why these differences lead to the use of different tools and techniques.

THE NATURE OF SERVICES

If you're a service provider, at one time or another, you've faced the issue of measuring the quality of what you produce. You've probably experienced a certain degree of frustration. How do you begin to determine whether a service encounter is a high-quality one? More importantly, how do you begin to quantify it?

You might begin by looking at the methods that are used to judge the quality of goods, but you'll soon run into problems. To begin with, goods can be inspected before they're shipped out to the customer. Quality can be checked *before* the goods leave the factory to see if they conform to specific standards.

For example, if you're manufacturing men's clothing, you can pick up a finished shirt and see if it's the right color and right size. You can check to see that both sleeves are the same length. If you make toasters, you can plug one in *before* you put it in the carton and see if it performs the way it's expected to. Does it burn the toast when set to "light?" Do the slices of bread pop up when finished?

On the other hand, if you're a provider of services, you can't go through this type of quality assurance drill. The services you produce can't be touched, tasted, or calibrated to standard before you deliver them to the customer.

Services are also highly variable. No matter how automated the process by which they are delivered, at some point, human intervention takes place. This intervention—the interaction between your customer and your employee—will be subject to the whims and unpredictability of human behavior. No matter how stringent your guidelines or how well-meaning your customer contact employees, let's face it: People are people! They're subject to stress, physical discomforts, personal problems—any and all of which can have a negative impact on their attitudes and behavior. Even the best employee may have an "off" day and react inappropriately to customer requests.

On top of this, there is variability between people: No two will react exactly the same way in the exact same situation—even if they say the exact same words, their tone of voice and expression will differ. They do not lend themselves to the traditional types of tools and techniques that are used to sample products and services and assure their quality before they are in the hands of the customer.

In fact, services can be considered to be transient. They are produced for, and immediately consumed by, the customer. This frequently means that you can't provide a service without a customer. Your ability to provide high-quality service to the customer is going to be highly dependent on your ability to interpret what the customer wants and to respond with the appropriate actions.

Customers are also subject to the same variability of human nature that afflicts service providers. They can be maddeningly imprecise or assign different meanings to the same request. For example, if you're a barber, what does "short on the sides and a trim on the top" mean? In a restaurant, what does an order for "rare" really mean? There are no commonly agreed on standards, no sets of international calibrations (as in engineering) to help you in such situations.

Finally, services can't be produced in advance and stockpiled or inventoried for periods of heavy demand. There is a finite period—a time window—in which you can offer your service or else lose the opportunity *and* the customer. If you're an instructor, openings that are not filled in the class you're teaching today cannot be filled months later. Seats on a particular airline flight cannot be sold once the plane takes off.

There is a greater danger when the demands for service cannot be met—lost customers. For example, if your customer experiences poor service during a period of heavy demand, such as in a crowded restaurant at lunch time, he or she will not be placated by your assurance that you can provide better service at off hours. Your customer wants the service here and now—and isn't likely to give you a second chance at a time that is more convenient for you.

HOW CUSTOMERS JUDGE THE QUALITY OF SERVICES

In Chapter 1, we discussed how a customer's judgment of service quality is affected by both process and outcome. *Process*, as you may recall, is *how* a customer is treated during the service interaction; *outcome* is the actual end result. For example, when you cash a check at your bank, receiving money is the outcome of this transaction; the way you are treated by the teller is the process. When you bring your car for routine maintenance, checking, tuning, and parts replacement constitute the outcome. The behavior of mechanics and the service manager in their interactions with you represents the process part of the transaction. For the latter service, in fact, the process may weigh most heavily in the service transaction. After all, most of us don't witness auto mechanics in action, and, even if we do, we may not have the technical knowledge to judge their expertise.

As different as these two examples are, however, there are certain basic elements that we would look for in each transaction, certain behaviors and results that would signal good or bad service. We can identify them by thinking back to times when we've had good service experiences and, unfortunately, the times when we've had bad ones.

As customers, we enter any service encounter with a series of expectations about what we expect to receive and how we expect to be treated. We then measure the performance of the service provider against our expectations. Where the service provider is lacking, where he or she fails to meet our expectations, there is a gap. The more gaps there are and the bigger the gaps, the more dissatisfied we will be. We will judge the service to be of poor quality. On the other hand, if the provider *exceeds* our expectations, we are likely to evaluate the service we receive as high in quality.

But on what specific elements of the service encounter do we center our expectations? To answer this question, we must think back to what distinguished our good experiences from our bad ones. Obviously, the end result had to be positive: The service had to be performed correctly (insofar as we were able to judge) and had to be

47

completed when we wanted it. But there were also many aspects of service that focused on the process part of the transaction. What kind of treatment did we receive? Were we seen as individuals or just as a number or account? Did the service provider really listen to us, communicate clearly, and react to us in a courteous manner? Or were we ignored, treated in a patronizing manner, and subjected to jargon? These same types of questions—and their answers—guided some pioneering research done in this area by Len Berry, Parsu Parasuraman, and Valarie Zeithaml.

THE DIMENSIONS OF SERVICE QUALITY

The research explored what customers look for in services and how they evaluate the service. We'll describe each dimension so that you can understand in depth what is important to customers.

Reliability. This dimension is used to describe a service company's ability to provide the services it promises dependably and accurately. In other words, does the company do what it says it's going to do when it says it's going to do it? Is my claim paid in accordance with the provisions of my policy? Is my bill accurate?

Of all dimensions, this one is the most "outcome oriented" and the clearest to define and measure. It is more black and white and less subjective. It most closely parallels the traditional concept of quality: the absence of all defects and "doing things right the first time." A service company must meet a customer's expectations for reliability if it is to compete effectively in the marketplace.

Responsiveness. This dimension reflects a service organization's willingness to help its customers and provide prompt service. As such, it focuses on both the process of service delivery as well as an easily measurable outcome—time service. Does the company give me a specific time that the repairperson will show up? Does he or she arrive at that time? Is my account representative willing to take the time to answer my questions?

Once again, this is a dimension that has an outcome aspect—timeliness—that can be more objectively measured against customer expectations. However, the behavioral aspects—the willingness of employees to help the customer—are also important. They can help a service provider move beyond market acceptability and exceed customer expectations.

Courtesy. Politeness, respect, consideration, and friendliness are the customer expectations in this category, which is strictly process oriented. Does the ticket agent greet me with a smile? Does the information operator answer the phone in a pleasant tone of voice? Is the baggage handler rude? Does the waiter ignore me?

Because this dimension is behavioral, it is more difficult to measure and evaluate. It is highly subjective, as it depends on the value judgments of individual customers.

Competence. Service organization employees must have the skills and knowledge needed to perform the service customers desire. When I call for information, can the customer contact person answer my questions? Does the plumber appear to know what he or she is doing? Can the waiter explain the ingredients of an item on the menu if I ask about it?

This dimension is particularly important when the service being provided is complex or technical and the customer has to depend on the service provider for advice or counsel. It is closely allied to the next dimension, credibility.

~ *Credibility.* This dimension encompasses the trustworthiness, believability, and honesty of the service provider. Like the preceding dimension, it is particularly important when the service in question is complex or, as in financial services, sensitive.

Here, the customer asks questions about the reputation of the service provider. Is the brokerage firm known for its integrity? Does the salesperson exert pressure on me to buy? Does the mechanic guarantee his or her work? Do I have confidence in the advice my dentist just gave me?

~ *Security.* This dimension can be characterized as freedom from danger, risk, or doubt. Do I feel secure when I use my bank's automatic teller machine? Can I immediately report a stolen credit card?

~ *Accessibility.* Approachability and ease of contact are included in this dimension. How accessible is the service? Does the company offer me a toll-free number? Can I place my order 24 hours a day? Can I call my utility company during nonbusiness hours? Does the auto insurer have a drive-in claim facility? Will the store deliver at night or on weekends?

~ *Communication.* This dimension refers to the service organization's verbal interactions with customers. It includes the ability to keep customers informed, to use language they can understand, and to listen to them. For example, does the airline let me know if there's a change in departure gates? Does the furniture store call me to let me know the delivery truck broke down while on route to my home? Does the doctor explain the diagnosis in understandable terms? Does the claims adjuster really listen to my questions, or am I given an answer to a question I didn't ask?

~ *Understanding the Customer.* This calls for a service organization to make the effort to know customers and their needs. Am I recognized at the restaurant that I frequently patronize? Does the insurance agent understand my financial objectives and recommend appropriate products?

~ *Tangibles.* Given that we've spent a great deal of time describing the intangible nature of services, it may seem strange that we include tangibles in our discussion of service quality dimensions. However, every service has a concrete aspect to it: the physical facilities in which it is provided, the equipment used in its delivery, the appearance of the personnel who serve the customer, the materials used to communicate. Tangibles include everything that the customer can see, hear, taste, or smell.

While this dimension is generally given the least weight by customers of a service, it cannot be ignored. Its importance will vary from customer to customer. For example, the roominess of an airplane seat may be more important to a 7-foot-tall basketball player than it is to a child. The type of service will also have a bearing on the importance that tangibles play in the evaluation of a service. The physical surroundings will generally mean more to us when we are patrons in a restaurant than when we are customers at a bank.

The 10 dimensions just described present a picture of what the customer looks for in a service interaction. You probably would not argue about their ability to summarize and encompass service excellence, for your own experience as a customer will confirm their importance. If you look back to good service experiences, you'll find that all or most of these dimensions were present. Conversely, bad service experiences are marked by the absence of these dimensions.

The following story shows how a bad situation can be made worse by a service provider.

ந. ந.

"Good Customer Service Is No Accident"

Kate, our well-traveled consultant, returned to her office in a bad mood, having suffered through a minor automobile accident while on assignment. She had been driving in the right lane when a panel truck in the middle lane attempted a lane change—right through her left rear fender. Fortunately, no one was hurt, and her car sustained only minor damage. A hasty trip to the auto repair shop for an estimate of the work needed had confirmed that this would be a relatively minor repair.

What had really upset her was the treatment she received from the claims adjuster when she telephoned the other driver's insurance carrier.

"I was upset enough by the accident, and she treated me like a spoiled child," Kate complained, when relating the story to client John Allen. "I called to report the accident and to ask what steps I needed to take to make my claim. She was abrupt and patronizing, and I felt I was pulling the information out of her."

"Your claim isn't a big one," said John, "and she probably felt you were bothering her."

"That's right," said Kate, "my claim is a relatively small one and won't exceed the deductible I have with my own carrier. But I don't think that's an excuse for her behavior on the phone! I'm also not sure why she insisted that I had to have one of their inspectors look at the damage before they paid the claim. If this is so small a claim, why do they bother having an inspector look at it? While they schedule a meeting for an inspection, my mechanic could be fixing the damage."

"Kate, you need to be more realistic. You know how auto insurance companies are. Even though the other driver admitted it was his mistake, they'll probably try to prove you were partly responsible. Then, they'll schedule the inspection and cancel the appointment several times. No matter what happens, it will probably be at least a month before you get the car fixed. In the meantime, at least you're able to use the car. Just brace yourself for plenty of petty annoyances—things don't happen that fast in the wonderful world of car repair! Are they going to send you the forms?"

"Yes. The woman I spoke to said she would send them right out." Back at her own office, Kate pondered John's admonition. Well, he's certainly lowered my expectations, she thought. I don't want to burst a blood vessel, so I hereby resolve to be more patient.

A week later, she received the promised forms, dutifully completed them, and returned them to the insurance carrier the next day. However, when three weeks passed and Kate had received no acknowledgment, she decided to call the adjuster and check the status of the claim. She dreaded calling Jean Smith again, but felt it was necessary.

The phone rang so many times, it left Kate wondering whether anyone was "at home." Finally, someone picked up.

"Claims, Jean Smith."

"Hello, this is Kate Nova, claim number 9D0313-74644," said Kate, anticipating the inevitable first question. "I called you about a month ago reporting a minor accident on the Major Deegan Thruway. The forms came about a week later, and I completed and mailed them back right away. I'm calling to check on the status of my claim and to set up an appointment with an inspector." There was silence on the other end. "Hello? Hello? Is someone there?" said Kate, thinking she had been cut off.

"Yeah, I'm here, I'm just looking through my pile for your claim—hold on." Kate heard the loud sounds of papers being shuffled, heavy objects being dropped, and tearing noises. Several minutes went by, which seemed to Kate like hours.

"Now I have your claim," said the claim adjuster. "It's a small claim. Why are you calling?"

I thought I just told you that, thought Kate. Instead she said, "Well, I'm still driving a damaged vehicle, and I want to set up a time for it to be inspected so I can get the repairs done."

"Why aren't you dealing with your own carrier?"

Kate felt her composure slipping and her patience dropping to a new low. "We've been through this before, Ms. Smith. The owner of the other car, your policyholder, told me to call this in myself. He is not contesting it—it's his fault and no one was injured. I explained all this on the forms you sent."

Kate heard some papers being shuffled again. "I guess you did," replied Jean Smith. It occurred to Kate that this was the the first time she was looking at the forms. "Well, since this is a 'nuisance' claim, why haven't you had it repaired?" said Jean Smith, annoyance strongly coming through in her voice.

"Because *you* told me to wait for an inspection," said Kate, "something I thought was unnecessary a month ago, but you insisted on—and I don't like being referred to as a nuisance!"

"I didn't tell you to wait for an inspection. We don't bother sending inspectors out for small claims. It's too expensive. You misunderstood me," said Jean.

Whatever patience Kate had left her at that moment. "Look, I don't like your attitude and I really want some answers. Can't I speak to someone else?"

"I don't like your attitude either," replied the claims adjuster to an incredulous Kate.

"Fine, I can see that talking to you is not getting me anywhere," responded Kate, and she hung up.

Well, thought Kate, being a consultant has its advantages. One of them is that you get to meet many people—especially management people. I know several key managers at that casualty company. I'll call that vice president of customer service I met while on that consulting project last year. Her handy telephone book supplied the number. A few minutes later, she was on the phone with Jim Green.

After a complete run-through of her conversations with Jean Smith, Kate concluded, "Jim, I hate doing this, going around people, but I really feel upset. What is your policy regarding claims such as mine—do I need an inspection?"

"Not at all, Kate. You're right to be upset. I can appreciate how you feel, having gone through this kind of situation myself. No, we don't require inspections for your type of claim. Please go ahead and have your mechanic do the work. In the meantime, I'll get a hold of the file and personally see to it that the claim is processed today."

"Thanks so much, Jim. I realize that I'm not your customer and maybe I shouldn't expect the same kind of treatment."

"The fact that you aren't currently insured with us is all the more reason why we should be trying to please you—if we impress you with our service you might buy a policy from us next time you're up for renewal. Kate, we try to get the message out to everyone, but communicating about customer service is very tough. Certainly, it's harder that I thought it would be when we started. Creating a policy is one thing—getting everyone to carry it out is another."

"Thank you for your help, and it's been nice talking to you, Jim. I hope we see each other in Boston at the customer service conference next month."

"Yes, I plan to be there. I'll be back to you no later than tomorrow morning. And again, I'm sorry this happened."

"Good-bye."

That night, Kate's message machine at home chirped, "Hi, Kate. It's Jim Green. The check went out to you this afternoon. If you have any further problems, please call me."

Well, Jim certainly kept his promise, thought Kate. He got back to me, and it looks like I'm getting my money without further hassle. I thought they might make me produce a statement from my witness. Now, let's see how long it takes to get my check, she thought cynically.

To her surprise, the very next day Kate received the check in the mail. She was impressed at how quickly Jim had responded to her complaint. What really pleased her, however, was to see Jean Smith's authorizing signature on the check. She had a vision of Jean Smith receiving a call from vice president Jim Green to process her claim. Kate smiled to herself. Yes, she thought, revenge *can* be sweet.

๊๏ ๊๏

Sadly, many of us can relate to Kate's encounter with the claims adjuster. We've all had experiences that we expect in advance will be unpleasant. We brace ourselves for the worst. However, even though Kate's expectations were low, Jean Smith's performance was even lower.

Kate was lucky—she knew where to go to get help, and she was dealing with a company whose management wanted to provide high-quality service. Jim Green understood that proper handling of telephone calls was not only the key to retaining existing customers but also a means of attracting new ones. In dealing with Kate, Jim concentrated first on *process*, listening to her concerns and reassuring her. Then, he turned his attention to her desired *outcome*, namely, settling the claim. He saw Kate not as a nuisance but as a potential policyholder. Unfortunately, he ran into harsh reality when converting this belief into action.

As service providers, we face the same challenge that Jim Green faced: How can we ensure that these dimensions characterize the service we offer to our customers? Just as when we discussed creating an effective organization in Chapter 3, the key is measurement. After all, we can only manage what we can measure. But measuring our success by how the customer perceives we are achieving the dimensions can be a challenge. Some of the dimensions are more objective and focus on outcomes, such as accuracy and time service. But the behavioral elements are evaluated more subjectively.

How do we set about the process of measuring both the process and outcome elements? Whose judgment should we rely on? The standard must and should be the customer. But how do we get the customer to quantify service quality dimensions in a way that allows us to track our performance and helps us to identify opportunities for improvement?

The answer can be found in an approach to measurement known as SERVQUAL. In Chapter 5, we will describe the SERVQUAL questionnaire and how it can be applied to any service.

QUESTIONS FOR DISCUSSION

"Good Customer Service Is No Accident"

1. Which of the 10 dimensions was Kate Nova looking for in her conversations with Jean Smith? Which did Jean Smith exhibit?

2. Describe how Kate's expectations changed throughout the course of the story. What influenced her?

3. What steps did Jim Green take to correct a bad situation? Which service quality dimensions did he practice? How effective was he?

4. What actions would you recommend Jim Green take to prevent recurrence of this type of problem in the future?

MEASURING QUALITY FROM THE CUSTOMER'S PERSPECTIVE: CUSTOMER FEEDBACK TOOLS AND TECHNIQUES

In Chapter 4, we discussed how services differ from products and how this affects the way customers evaluate service quality. We identified the basic dimensions that customers look for in a service interaction—dimensions that reflect both *how* the customer is treated and *what* the customer ultimately receives.

The questions service providers usually ask at this point are as follows: How can this concept be used to quantify service quality? How can it be translated into action plans designed to bring about improvements in service quality? In other words, can you really measure service quality from the customer's viewpoint? We will answer these questions by describing the SERVQUAL instrument and then by illustrating how it can be applied to any type of service.*

---

*"SERVQUAL: A Multiple-Item Scale for Measuring Customer Perceptions of Service Quality," A. Parasuraman, Valarie A. Zeithaml, Leonard L. Berry, Report No. 86–108, Cambridge, MA: Marketing Science Institute, 1986.

THE SERVQUAL INSTRUMENT

SERVQUAL is a survey instrument developed by the Berry-Parasuraman-Zeithaml research team. It consists of two major parts: one to capture customer expectations for a given type of service, and the second for customer perceptions of experiences with a given service provider.

The parts consist of parallel statements. Each statement focuses on an aspect of one of the dimensions of service quality and has a response scale ranging from 1 to 7. The scale is used by a customer to indicate the degree to which he or she agrees or disagrees with the statement.

Responses to the expectation and perception statements are compared. A score is computed by subtracting the expectation response from the perception response. *If the expectation response is higher than the perception response, the score will be negative.* A negative score indicates the existence of a service quality gap: Customers are not having their expectations met by the service provider. On the other hand, *a positive score is the result of performance that exceeds customer expectations.* A positive scores indicate an area of strength and can represent a competitive advantage for the service provider.

Here's how this would work. We'll use the Quick Stop Dry Cleaners as our example. The owner wants to measure the performance of his store, as it's perceived by the customer. We'll begin with a statement pair that describes an aspect of the reliability dimension:

Part I: *Expectation Statement*
           When excellent dry cleaners promise to have clothing cleaned by a certain time, they will do so.

Part II: *Perception Statement*
           When the Quick Stop Dry Cleaners promises to have my clothing cleaned by a certain time, it does so.

The response scale for each of these items would range from "1, strongly disagree," to "7, strongly agree." Suppose a customer responds with a "7" in the first part of the questionnaire, indicating the strongest possible expectation for having clothing cleaned by the promised time. If the same customer responds with a "5" on the matching perception statement, the score can be computed as follows: 5 (perception) −7 (expectation) = −2. For this particular customer, Quick Stop Dry Cleaners has not met reliability expectations—specifically, those that have to do with meeting promises on cleaning times.

DEVELOPING A CUSTOMIZED QUESTIONNAIRE

While the SERVQUAL approach is not especially complex, there are several pointers to keep in mind when developing and applying it.

Expectations Should Be Generic for the Service. This means that all the expectation statements in the first part of the questionnaire should deal with the service in general, not the service offered by a particular provider. In our example, note that the expectation statement dealt with expectations for any excellent dry cleaner, not for Quick Stop Dry Cleaners: "When *excellent dry cleaners* promise to have clothing cleaned by a certain time, they will do so."

Perceptions Should Be Specific for the Service Provider. This means the statement covering perceived performance should use the name of the service organization that is being evaluated. In our illustration, the perception statement specifically mentioned the vendor: "When *Quick Stop Dry Cleaners* promises to have my clothing cleaned by a certain time, it does so."

Only One Service Aspect Should Be Covered by a Statement. If more than one aspect is covered, the customer's attempts to respond may be frustrated. For example, consider the following statement: "The people who work in a dry cleaners should be courteous and knowledgeable about stain removal."

As respondent, I may want to assign a "4" rating to the courtesy part of the statement, but I may feel more strongly about the knowledgeability part of the statement and want to assign a top rating of "7" to that. How do I respond—with a "5" or a "6"? Neither mark would reflect my true expectations.

Even more important, suppose I perceive the people who work at Quick Stop Dry Cleaners to be extremely courteous ("7") but unable to remove stains from my clothing ("1"). I would find agreeing with that statement frustrating. The solution, then, is to ensure that separate aspects are covered by separate statements.

Expectation and Perception Statement Should Have Parallel Wording. This allows the customer to compare apples to apples, not to oranges. Consider the following pair of statements, which focus on reliability: "Employees who work for excellent dry cleaners prepare accurate bills." "Employees at Quick Stop Dry Cleaners do not make errors."

The people at Quick Stop Dry Cleaners may prepare accurate bills, but there is more to errorless service than computing charges, for example, delivering clothing to the right address. Again, as a respondent, my ratings on these two statements may be reflecting two very different types of experience. Thus, to ensure comparability, the statement pairs must have parallel wording.

Statements Should Describe Unambiguous Behaviors. This means that statements should be as precise as possible and should not include vague terms. It is especially important to use words that describe or define the dimension, rather than the dimension itself. This helps to ensure that customers are interpreting the statements in a uniform way. For example, the statement "Excellent dry cleaners are responsive" can be interpreted by different customers in different ways: being open at times that are convenient for the customer, offering same-day service, the willingness of service employees to go out of their way to help the customer, and so on.

Likewise, the statement "Quick Stop Dry Cleaners employees take a precise approach to their work" can have different meanings for different customers: opening on time, making correct deliveries, never ruining one's clothes, and having items ready at the promised time.

Statements Should Focus on Positive Aspects of Service. By this, we mean that the statement pairs should emphasize the desired service in positive terms. If this is not done, computing gap scores becomes complex.

Let's contrast the responses we would give to the same aspect of a dimension when the wording shifts from positive to negative. For example, *accessibility* is the dimension of service quality that relates to ease of contact. The following is a positive perception statement that reflects this dimension: "Quick Stop Dry Cleaners has operating hours that are convenient for its customers." I may strongly agree with this statement and give Quick Stop Dry Cleaners a "7." On the other hand, a *negative* perception statement would be as follows: "Quick Stop Dry Cleaners doesn't have operating hours that are convenient for its customers." If I have the same perception—I strongly believe that Quick Stop Dry Cleaners offers convenient operating hours for its customers—I would strongly disagree with the negative statement. Thus, my response here would be a "1."

As the example illustrates, the response scales for positive and negative statements are inversely related. This adds an extra degree of complexity in computing gap scores. Thus, to avoid an additional—and unnecessary—step, we recommend that all statements be phrased in positive terms.

STAGES IN SURVEY DEVELOPMENT

We've just covered general points to keep in mind as you go about developing a SERVQUAL type of instrument. However, you still have to develop the specific content—the actual wording of statements that describe the service you want to evaluate in terms of the dimensions of service quality. To do this, you will go through several stages of survey development, including collection of qualitative data, analysis of data to identify service features by dimension, development of statement pairs, refining the instrument, and testing and implementation.

Collection of Qualitative Data. To identify the aspects of service dimensions that are to be evaluated, it's important to go to those who know the most about the subject: your customers and the people who directly serve them, your customer contact employees. One way of doing this is by collecting qualitative data through the use of focus groups.

While there are market research specialists who are extremely skilled at conducting focus groups, an organization does not have to hire a professional to use this technique. A person who practices good facilitation skills can effectively fulfill the role of focus group leader. The key is to use open-ended questions and to remain neutral.

A good way to begin is to ask members of the customer group to describe experiences they've had with a given service—both good and bad. What influenced their judgments? What made the experience memorably good or memorably bad? While professional focus group facilitators often use video- or audiotapes to record responses, a more informal leader can take notes on flip charts or use pad and pencil. The important point is to refrain from editing customer responses or interpreting them in line with one's own beliefs and biases.

A similar approach can be taken with focus groups of customer contact employees. In this instance, however, the questions should center on what customers most frequently look for. What do customers ask for? What do customers seem to appreciate most? What do customers dislike most? Once again, both positive and negative anecdotes should be sought, with the focus group facilitator using open-ended questions and remaining neutral. To encourage open discussion, it may also be better to have someone other than the boss serve in the facilitator's role. After all, employees tend to shield the boss from less successful service encounters!

Analysis of Qualitative Data. Once you have all the focus group input from your customers and contact employees, it is time to try to make some sense of it. After all, focus group data do not spring forth neatly organized. Members will offer their opinions and anecdotes in no particular order; they will talk about several different aspects of quality in the same sentence.

For example, consider a few of the statements about dry cleaning service made by members of a customer focus group.

1. "You know you're dealing with a classy dry cleaning service if they carefully look over the garments you bring in and tell you if they see a loose button or check the pockets. It's even better when they sew the button on without being asked."

2. "I really don't like to be invisible to the person who takes my clothes. When I enter the store, they should greet me, treat me like a human being, and work quickly to get me out."

3. "It's nice if the dry cleaner you deal with has special services, like picking up clothing from your home and delivering it to you or sewing tears and buttons."

4. "I've been going to the same dry cleaners for years. They basically do a pretty good job and they're right near home, but what irks me is that the owner still asks me my name and how to spell it when he fills out the receipt. You'd think he would know by now."

Begin by breaking down statements that cover more than one aspect of service. In the previous example, statement 2 refers both to personalized treatment ("...[not] invisible...greet me...treat me like a human being...") and prompt service ("...work quickly to get me out"). After splitting statements where necessary, begin grouping similar statements or characteristics together: Are there certain behaviors that recur in the stories and anecdotes your focus group members have given you?

In statements 1 and 3, the customers mention dry cleaners who take an extra step to check clothing and make minor repairs ("...see a loose button or check the pockets...even better when they sew the button on without being asked...special services like...sewing tears and buttons"). Statements 2 and 4 both place a premium on personalized, individual treatment.

Once similar statements have been grouped together, consult the list of 10 dimensions in Chapter 4 and review their definitions. Can you relate your data to the dimensions of service quality? Are all dimensions covered? For example, statements that relate to personalized treatment reflect the dimensions of courtesy and understanding the customer. The speed with which the customer is attended to (part of statement 2) reflects the dimension of responsiveness. The desire for special services—taking the extra step—can be seen as reflections of reliability and of competence.

Use the list of 10 dimensions as a checklist. While it is not necessary to cover them all—some may be more important for a given service than others—be sure that none has been inadvertently omitted.

Development of Statement Pairs. Once you've grouped your data by dimension, begin to convert the dimension data into positive statements concerning the service in general.

For example, with dry cleaning vendors, the following expectation statements have been developed based on the focus group data presented above:

"It is important to me that the person I deal with at the dry cleaners does not let customers wait in line a long time."

"I expect a dry cleaners to sew loose buttons on garments brought in to be cleaned."

"It is important that the person I deal with at the dry cleaners is polite."

"It is important that a dry cleaner offer an optional home delivery service."

The matching *perception* statements for the Quick Stop Dry Cleaners are:

"The people I deal with at the Quick Stop Dry Cleaners do not let customers wait in line a long time."

"Quick Stop Dry Cleaners sews loose buttons on garments brought in to be cleaned."

"The people I deal with at the Quick Stop Dry Cleaners are polite."

"Quick Stop Dry Cleaners offers optional home delivery service."

Note that the statements are parallel and follow the exact order, whether they focus on expectation or perception.

Refining the Instrument. Once the statement pairs have been developed, it is time to transform them into a survey instrument that can be tested and further refined.

The instrument will have at least two sections. The first consists of all expectation statements; the second, all the perception statements. As we have already pointed out, the perception statements should be in the same order as their matching expectation statements. This does not mean, however, that all the statements designed to measure a given dimension (e.g., responsiveness) should be grouped together. In fact, it's a good idea to alternate differ dimensions when developing the order statements.

The information derived from comparing the statement scores from the first and second sections is sufficient to provide a starting point for improving service quality. However, a third section can be added to the instrument that will help the service organization to set priorities for its action plan. This section calls for customers to rate the importance of each dimension.

For the Quick Stop Dry Cleaners, section 3 might read as follows:

"Listed below are features that pertain to dry cleaning stores and the services they offer. We'd like to know how important each feature is to you when you are judging the quality of service you receive from a dry cleaning company. Please allocate a total of 100 points among the features according to how important each feature is to you. The more important a feature is to you, the more points you should assign to it. But, please be careful to ensure that the points you allocate to the features add up to 100."

Following the listing would be statements that define each of the dimensions of service quality for the service being evaluated, along with a line on which point allocations can be entered. For example, the statements for tangibles, reliability, and responsiveness in section 3 of Quick Stop Dry Cleaners' survey are as follows:

The appearance of the dry cleaning company's physical facilities, equipment, personnel, and communications material.   _____ points

The dry cleaning company's ability to perform its promised service dependably and accurately.   _____ points

The dry cleaning company's willingness to help customers and provide prompt service.   _____ points

Allocating points among the 10 dimension categories, however, can be burdensome for survey respondents. How can we make it easier for them to differentiate between dimensions? It would be helpful if we had a more manageable list of dimensions. Fortunately, further research by the Berry-Parasuraman-Zeithaml team has led to the development of two broader dimensions that cover the seven remaining dimensions from our original list of 10: competence, courtesy, credibility, security, accessibility, communication, and understanding the customer. Various statistical analyses indicated considerable correlation among the seven dimensions listed above and led to their consolidation under the dimensions of *assurance* and *empathy*.

Assurance is the knowledge and courtesy of employees and their ability to convey trust and confidence. Empathy is the caring, individualized attention the firm provides its customers.

While we believe that the original 10 dimensions should guide the development of statement pairs, we recommend that the five dimensions be used in section 3 of the instrument. This reduces the number of choices customers face when we ask them to identify the most important features of the service and makes it more likely that we will receive meaningful data to analyze. This reduced number of choices will also be helpful in the later stages of our work, as we attempt to develop a baseline against which we can measure our future progress.

61

TESTING AND IMPLEMENTATION

As with any new product or service, a survey instrument should be tested before its wide-scale introduction. Customers, of course, are the best people for testing a customer survey. In conducting such a test, however, it is important to make sure that people participating in the test are representative of the customer base. For example, if Quick Stop Dry Cleaners offers a pickup and delivery service, its pilot test should include customers who use this service as well as those who bring their clothes into the store and pick them up themselves.

The test can be done with small groups of customers or one on one. The key consideration is time. There should be a sufficient amount of time for customers to complete the survey and then discuss it with the survey administrator. Customers should be encouraged to identify any problems they have in responding to the questions. Were the questions clear and unambiguous? Did customer have trouble responding to any statement? Did the questions cover all aspects of the service customers consider to be important?

Also, be alert to any feedback concerning the length of the questionnaire. Balance the desire to collect enough information to cover all the dimensions with the need to minimize the demands on customers' time. Remember, customers are likely to ignore any questionnaire that runs on for several pages or scores of items. Thus, survey items present a clear case of less is better than more. Our rule of thumb is to aim for no more than 25 pairs of expectation and perception statements.

Once you obtain customer feedback and do necessary refinement, it's time to tackle the questions that surround survey administration.

1. Will all customers receive our questionnaire?

2. If it is to be distributed only to a sample, how will that sample be selected?

3. Are there any special steps we can take to encourage many customers to complete and return the survey?

4. Who will receive the completed surveys?

5. Who will record the responses and compute the gap scores?

All of these questions need to be answered and decisions made before a single instrument is sent out.

In answering these questions, however, certain factors must be considered. The numbers of customers involved in testing and in the actual implementation and the responsibility for survey administration depend on the size of the organization and its customer base. Our general recommendation is that large organizations with large customer bases should call on experts to develop, test, and administer survey instruments. Research firms that specialize in this area are likely to have both the expertise to develop a professional instrument and the resources to collect the raw data and compute gap scores within a reasonable time frame. Assigning responsibility for data collection to an outside source also helps to safeguard data integrity and validity.

Smaller organizations can allow more informality in survey development and administration. Advanced techniques and data correlations are less important because such organizations may not have enough customers to provide a statistically significant sample. To evaluate the adequacy of their customer survey techniques, smaller organizations need to ask the following questions:

1. Have we taken sufficient steps to identify the major aspects of service quality that are important to our customers?

2. Have we related the customer information to the recognized dimensions of service quality?

3. Have we developed an instrument that enables us to compare customer expectations to customer perceptions of our performance?

4. Have we tested the instrument with customers?

5. Have we taken steps to ensure the adequacy and validity of the data we collect?

An organization that answers yes to all of these questions is then ready to distribute its survey.

INTERPRETING SURVEY RESULTS AND DEVELOPING ACTION PLANS

As customers begin to return their surveys, the need to collect and analyze data becomes acute. Once again, the resources devoted to this step will be determined by the size of the organization and the number of customers. Larger organizations may use an internal or external survey group; smaller organizations may use a personal computer spreadsheet or a pad, pencil, and calculator. Whatever the means used, however, the method of analysis remains the same. Gap scores are computed for both individual items and for clusters of dimensions, and scores for dimension clusters are weighted to reflect importance to customers.

Computing Gap Scores for Individual Items. As we indicated earlier, a service quality gap is computed by subtracting expectation responses from their matching perception responses. Let's assume the Quick Stop Dry Cleaners developed a survey instrument that has 12 pairs of statements. We'll focus on one of the statements that reflects reliability. The expectation and perception responses of 10 respondents appear in Table 5.1.

As Table 5.1 indicates, the gap score for this item is the sum of all the individual differences between matched expectation and perception responses divided by the number of customers who responded. In our example, the gap score for this item is −0.6. This process is followed for each statement pair.

Computing Gap Scores for Dimensions. You may develop a survey instrument that has more than one statement pair reflecting a given dimension. In such cases, the scores for the statements on a given dimension should be aggregated, and the

Table 5.1 Quick Stop Dry Cleaners Reliability Item

Responses from 10 Customers

Perception Score	Expectation Score	Difference
7	6	+1
6	7	−1
5	7	−2
5	6	−1
4	7	−3
6	7	−1
6	6	0
7	7	0
5	6	−1
7	5	+2
	Total	−6

Total Divided by Number of Respondents: −6 / 10 = −0.6

Gap Score: −0.6

average gap score for the dimension should be computed. For example, if three statements dealing with the reliability dimension have individual scores of −1.2, −0.6, and −0.9, the overall gap score for the reliability dimension is −0.9, the average of the three statement scores.

This clustering of scores by dimension is useful for two reasons. First, it establishes a quantitative baseline against which future progress can be tracked. It thus provides a way to gauge the success of efforts to improve service quality. Second, combining the gap scores for each dimension with customer importance ratings develops priorities for action.

Developing Weighted Gap Scores. Let's assume the Quick Stop Dry Cleaners has aggregated the item scores for statement pairs according to the dimensions they reflect. The gap scores for each of the five service quality dimensions is as follows:

Dimension	*Gap Score*
Tangibles	−0.1
Reliability	−1.0
Responsiveness	−1.3
Assurance	−1.2
Empathy	−0.4
Total	−4.0

The overall gap score is −0.8, the total of −4.0 divided by 5 (the number of service quality dimensions).

At first glance, the management of Quick Stop Dry Cleaners may decide it should direct its attention to responsiveness, which has the higher gap score. But its survey also had a section in which customers were asked to rank dimensions in terms of their importance, by allocating 100 points among the five dimensions. The results for 10 respondents are as follows:

Dimension	Point Allocated	Percent of Total	
Tangibles	50	5%	(0.05)
Reliability	300	30%	(0.30)
Responsiveness	200	20%	(0.20)
Assurance	300	30%	(0.30)
Empathy	150	15%	(0.15)
Total	1000	100 %	

These percentages can be used to compute weighted gap scores for each dimension. This is done by multiplying the gap score for the dimension by the weight given to that dimension by customers.

Dimension	Gap	Weight	Weighted Gap
Tangibles	−0.1	0.05	−0.005
Reliability	−1.0	0.30	−0.30
Responsiveness	−1.3	0.20	−0.26
Assurance	−1.2	0.30	−0.36
Empathy	−0.4	0.15	−0.06
			−0.985

This further analysis provides the owner of Quick Stop Dry Cleaners with additional information. First, the overall service quality gap is greater than it originally appeared to be. The original gap score was −0.8 (−4.0 divided by 5 dimensions); the weighted gap score is −0.985.

Second, the weighted scores would lead Quick Stop Dry Cleaners to reevaluate its priorities for action. At first glance, the unweighted scores would suggest that the gaps, in order of seriousness, are responsiveness (−1.3), assurance (−1.2) and reliability (−1.0). However, a look at the weighted score leads to a reordering of priorities: assurance (−0.36), reliability (−0.30), and responsiveness (−0.26).

IDENTIFYING OPPORTUNITIES FOR IMPROVEMENT

All the analyses in the world, however, will never produce improvements in customer satisfaction or service quality. Something has to happen or change as a result of all this work. The bottom line is action: identifying areas of service that need to be improved and developing the action plans to bring about increased customer satisfaction—to close the gap between customer expectations and perceptions.

A good starting point is the weighted gap score. Look at the dimension that has the biggest gap and then review the detailed items which make up the dimension, that is, the individual statement pair scores. Ask the following questions:

1. Which item or items under the dimension have the biggest gap?

2. What behaviors or processes are reflected in this item?

3. How do we currently operate?

The answers to these questions serve as background information for a problem-solving meeting by those responsible for providing the service.

Let's say that after reviewing weighted gap scores, Quick Stop Dry Cleaners identified assurance as the dimension most in need of immediate action. The statement within this dimension that has the highest (most negative) gap score has to do with customer contact people's knowledge of stain removal. Action plans can focus on several aspects: how to enhance employee knowledge of stains, how to improve stain removal during the cleaning process, how to communicate with customers about how the stains happened, and so on. Thus, the survey data provide practical, actionable input for problem-solving efforts to improve customer satisfaction.

Moreover, surveying customers forms part of any organization's continuous improvement efforts. You don't do a survey once and think that you now understand your customers needs and expectations forever. Customer's expectations, like anything else, change over time. In fact, if you improve, they'll expect more the next time they come to you for service! You should add regularly scheduled surveys to your service quality plans. Of course, large organizations with numerous customers should survey more often than small organizations. However, we recommend that any organization survey its customers at least once a year.

SUMMARY

The SERVQUAL approach provides service organizations with a way of quantifying service quality from the customers' perspective. With SERVQUAL, you can identify those features of the service that your customers use in judging its quality. You can also learn which of the various service dimensions are most important to your customers. Finally, you can use SERVQUAL to develop an objective baseline measure of your service and to identify those areas that need improvement. This approach represents the best current thinking about the complex reality of customer opinions about your service.

Surveying your customers periodically and keeping a record of gap scores starts you on the road to improved customer satisfaction and provides you with valuable insights into your quality efforts over time. Understanding your customer's view of quality and identifying customer-perceived gaps in your service provide the most important starting points for action.

However, as we said earlier, customers have a somewhat limited view. They judge your organization as a whole by the contact they have with one or two members. Like members of an audience, they are uninterested in the internal processes you go through to serve them or the numbers of people behind the stage who are essential to your performance. They evaluate you on end results. What did they receive? How were they treated? Did your performance meet their expectations?

Likewise, traditional efforts to examine and improve the process by which we provide service to the customer often address immediate problems and will quickly have a positive impact on customer satisfaction. However, traditional quality efforts may run out of steam. Members of the organization experience fewer big breakthroughs in service improvement and may become discouraged by a slowdown in the rate of progress they are making.

Why does this happen? All too often, it is because the original issues addressed scratch the surface of the organization. They may not get at the heart of the matter, namely, the underlying attitudes, practices, and management approaches ingrained in the organization's culture that can contribute to poor quality. When viewed from inside the organization, these issues display themselves as a series of internal gaps that stand in the way of meeting customer expectations. In Chapter 6, we will identify these gaps and begin to discuss the actions we can take to address them.

QUESTIONS FOR DISCUSSION

The Quick Stop Dry Cleaners conducted a customer focus group to identify the aspects of dry cleaning service that their customers use to evaluate the quality of their service. Quotes from the focus group members appear below:

"I want the person at the dry cleaner to know more about stain removal than I do—and I don't want them to lie about whether they can get a stain out. I would rather they tell me up front that they can't do something than raise my hopes."

"I really don't like to be invisible to the person who takes my clothes. When I enter the store, they should greet me, treat me like a human being, and work quickly to get me out."

"I hate when you're in the dry cleaners and they have long personal conversations with another customer ahead of you, and you're there waiting in line."

"The clothes should be cleaner when I pick them up than when I brought them in."

"Sure they should be cleaner, but they should also be well-pressed. Pressing something well is an art."

"I don't mind waiting a reasonable amount of time for my items, if they are cleaned and pressed well when I get them back. I think three days is a reasonable amount of time."

"I think price is important, but what's more important is that they salvage a garment that is badly stained. I'm willing to pay a little more if I had confidence in the people who do the cleaning."

"The people who take the clothes are sometimes not the ones who do the cleaning, so I expect them to really listen to me and tell the others what I want done or what caused the stain—things like that. I would like to feel that they really understood me. Frequently, they act as if they were unconscious."

"Once, I was given someone else's bedspread when I went to pick up my order. I had to lug the thing back after unwrapping it. Not only that, I had to wait three or four days before they could track down the mistake and give me my bed-spread. That was the worst thing a dry cleaners ever did to me."

"You know you're dealing with a classy dry cleaning service if they carefully look over the garments you bring in and tell you if they see a loose button or check the pockets. It's even better when they sew the button on without being asked."

"It's nice if the dry cleaner you deal with has special services, like pickup and home delivery or sewing tears and buttons."

"I expect that as their customer, when I have an emergency, like I need a suit in 24 hours, they can do that for me with no hassles. I like a place I can depend on, especially because I travel for my job."

"I've been going to the same dry cleaner for years. They basically do a pretty good job and they're right near home, but what really irks me is that the owner still asks me my name and how to spell it when he fills out the receipt. You'd think he'd know that by now."

"I think that dry cleaners should encourage recycling of hangers. It would be better for the environment. I don't like to use wire hangers in my closets and always bring them back to the store to be used over again. I get a thank you, but I'd like a small discount on my bill even better. We have to be more conscious of the garbage we create."

"I think the way a dry cleaning service wraps the items you bring in is important because sometimes I store them away just as they are given to me."

1. Which service quality dimensions can be related to each of the customer comments?

2. Do comments cover more than one dimension? If so, break them into their component parts.

3. Group similar comments together and develop expectation and perception pairs for each.

4. Develop a plan of action for Quick Stop Dry Cleaners. What "next steps" would you recommend?

INVOLVING EMPLOYEES IN CUSTOMER SERVICE IMPROVEMENT EFFORTS

The first part of our book focused on how to develop a strong customer orientation and discover ways to create more effective service organizations. The organizational self-assessment, the customer-provider work flow, and the use of customer surveys, such as SERVQUAL, are all helpful techniques, because they provide structured approaches to developing this type of orientation. The use of these techniques forces an organization to face its fundamental reason for being and the basis of its existence—namely, attracting new customers and retaining existing ones. Moreover, the techniques enable all organizational members to understand how their work— for better or for worse—has an impact on customer satisfaction.

The real power of these techniques, however, rests in their ability to involve all employees in the process of improving customer service. The service provided to an external customer is the direct result of a series of interactions between internal customers and providers. By improving these internal interactions, service delivery to the ultimate customer will also improve. It then follows that the people who are best able to identify opportunities for internal service improvements are the same people who will implement these improvements. This is because they are part of the customer-provider work flow.

It's obviously a good decision to involve the people who contribute to the creation and provision of a service in both the self-assessment process and the development of customer-provider work flows. After all, they live with the process every day and know the details. This makes them a powerful source of improvement ideas. It also makes sense to bring together those who have different perspectives because it enhances the overall creativity of the ideas and solutions. Feedback of the results of a customer survey can be a powerful motivator for teamwork.

71

In addition, by involving those responsible for implementing changes, obstacles are more likely to be identified up front. The obstacles are then faced and overcome *before* the change is implemented. This saves critical resources, such as time and money. A final important advantage gained is that people who have a say in developing solutions are more likely to be committed to making them succeed.

THE BARRIERS TO PARTICIPATION

For the reasons just discussed, it's easy to see the benefits of involving employees in efforts to understand customer needs and improve service delivery. What's not so easy is getting individuals from different areas to work together effectively in a group setting. All the techniques in the world, all the data gathered and analyzed will do little good for the customer if people don't know how to work together.

What can go wrong? The first problem is that we are bringing together people from diverse parts of the organization who often have conflicting goals. In many instances, they have not thought of each other as customers and providers. They may have a history of adversarial relationships. This creates an atmosphere of distrust, dominated by hidden agendas.

Even in situations in which all members have a good understanding of their relationship to the ultimate customer and agree on overall goals, they may disagree on the means to achieve them. This situation creates a further strain, when different levels within the organizational hierarchy are represented. Those lower on the totem pole are likely to be ill at ease and reluctant to identify problems or offer suggestions.

Let's now meet a group of employees who, on the surface, do not seem to suffer from the problems that we've just identified. The employees have a good understanding of who their customers are and share the same goal. They report to the same boss and represent just one function—management training. Let's look at how effective they are at working together to improve what they deliver to their customers.

₨ ₨

"If You Don't to Want Listen, Why Ask?"

"We've already waited ten minutes—can't we get started?" asked Alicia. "I'm expecting a call from the copy editor. You know how busy she is. If I miss her, we'll end up playing telephone tag all day."

"But Joan called this meeting," Marla pointed out. "We really have to wait for her."

"Unless Roger will let us in on the big mystery," said Jeff. "C'mon, Roger, shoot. Why are we all here?"

The small talk that had filled the room moments ago trailed off as all eyes turned to Roger Johnson, their project manager. As Roger looked around at the members of his team—Marla, Alicia, Jeff, Peter, and Sam—his thoughts turned to the project they were working on. I couldn't ask for a better bunch of trainers, he thought.

The new program they were developing about managing change was probably the best seminar his unit had ever developed. Feedback from the managers who'd participated in the pilot was overwhelmingly positive. In fact, upper management was pressing his team to bring the training program on line as quickly as possible. For the past two months, they'd been putting in many extra hours.

"Well, folks, we've all read that new management book. Joan wants to talk about it this morning. She feels our customers would really benefit from the ideas in it and enjoy the way they are presented."

"I liked it, too. It was fun to read," offered Jeff. "That's saying a great deal for a book on management principles."

"I had the same feeling," Marla added. "It was simple to understand, but the ideas were practical and had substance. I think our customers would like it."

"Then we should definitely let our participants know about it," said Alicia.

At that point, Roger's boss, Joan Janson, walked into the small conference room. Six pairs of eyes now focused on her. As she moved to her favorite chair at the far end of the room, she asked Roger, "Have you started to discuss how we can use the book?"

"We were all agreeing how good it is when you walked in, Joan."

"Great! Then it shouldn't take too much time to agree on a plan of action."

Alicia, thinking about her call to the copy editor, perked up noticeably. "How much time do you think we'll take, Joan?"

"I think half an hour will do."

"Fine."

Joan looked around the table and began.

"I'd first like to point out what I see as the positive effects this addition would make to our program. The book deals with complex ideas in a straightforward manner, something much appreciated by today's managers." Heads around the table nodded in agreement.

She continued. "By incorporating this new material into our current design, the managers—our customers—will benefit directly."

"Wait a minute, Joan," Alicia broke in. "What exactly does 'incorporating this material into our design' mean to you? If you're thinking of any changes in the lesson plans, count me out!"

Peter spoke up, trying to soothe Alicia. "Don't get so excited, Al. There are other ways of offering this material to the managers. We can just give them a copy of the book as a gift at the end of our sessions. And I'd bet the author would appreciate that, too!"

"What's good for the author should be of little concern to us. I'm more concerned about our customers," Joan shot back.

"I know you won't like this, but what about giving the book to the participants *before* the seminar, as prereading?" offered Marla.

Roger shook his head, rolling his eyes in disbelief. "Marla, your middle name must be Pollyanna! You really think the people who come to our program will actually read what we mail to them before the seminar? You have enough experience to know that never happens!" How naive can she get? he thought.

"Besides, giving the book out as supplementary reading lessens its importance," said Joan.

Sam saw how uncomfortable Marla was and tried another approach.

"I agree that the new material is valuable, well written, and even entertaining. But the concepts are already covered in our current design. We wouldn't just be adding material, we'd have to delete the areas where we already cover these ideas, so the training wouldn't be redundant."

"Sam's right," said Alicia. "We're already behind schedule, and I'm worried that the extension we were granted will not be enough—and that's without major modifications to the design."

"Afraid of working some overtime?" joked Roger.

Alicia snapped back. "Roger, I know you meant to be funny, but at this point, I'm too tired to laugh. I've been getting here early, leaving late, and working weekends for two months, and I know I'm not the only one.

"I'm sorry, Alicia, you're right," said Roger. "With some of the lesson plans already at the printers, changing the design now would mean resetting some of the print, moving the artwork—heaven knows what else. It would certainly add to our costs."

Jeff broke in, "Well, I'm glad someone finally mentioned money around here. What about the budget? We might not even have the money to buy several hundred copies of the book, let alone have money to start redesigning."

"I don't understand your attitude," said Joan in frustration. "You all agree this is great material, but you refuse to make it part of the seminar's design."

"Look, Joan, we've already given some very strong reasons why that's not feasible. If this book had come out six months ago, I think we would all feel differently," said Sam.

"Let's go over your objections again," said Joan.

"OK," said Roger, glancing at the notepad before him. "Basically, what it boils down to is that changing the current design would take too much time at this stage of development."

"Wait a minute, Roger," said Alicia. "I've got more than that one point in my notes. You didn't write down everyone's ideas. What about the increase in expenses if we change the design now?"

"And I can think of another argument against it," said Peter. "If we try to go back and change things now, chances are the quality of the program would suffer. With our tight deadline, we wouldn't catch all of the duplications and we'd probably leave some things out. Sheer exhaustion gets to you after a while!"

"And we'd have to pilot test again, wouldn't we?" Sam added.

Joan stood up. "This is getting us nowhere. I think we should forget about the book for now."

As the rest of the team left the room, Roger and Joan exchanged glances. When they were alone, Joan looked discouraged.

"That was a terrible meeting," she said. "I hate to say it, but I can see why the old-timers say participative management is a waste of time. Everyone was so emotional and unwilling to listen. I'm sure that if they'd really listened, they would have seen the merit of incorporating the new material. Now, our new training program isn't going to be as good as it could have been."

Roger remembered the first time he put on ice skates when he was a child. No matter how careful he was, he knew with certainty that he was going to fall.

The only question was how badly he would be bruised. He felt that same sense of certainty now.

"Joan, the program will be fine. We've already pilot tested it and it works."

"But they don't want to add the new material."

"Joan, if you felt that strongly about the new material and wanted the team to modify the seminar design, you could have ordered them to do it. After all, you are the boss."

"But it's not a question of being the boss, it's the right thing to do. I felt sure they would see that."

"That may be true, but wasn't the purpose of the meeting to explore the idea of adding new material, not necessarily to get a rubber stamp of your idea? I think that some of their concerns couldn't be overcome without adding significant time to our target date."

As she and Roger walked to their offices, Joan said, "I still think the people who attend the seminar would have enjoyed the book.

Roger just shrugged. "The team suggested giving the book to the participants as a gift," he pointed out.

"But that's not the best approach."

"It was an alternative. And, sometimes, when you ask a question, you don't always get the answer you want."

&a &a

At one time or another, we've all experienced a meeting like the one the training team just suffered through. We sincerely try to foster participation. We bring together those we feel are the best ones to tackle a particular issue. Then, we find that our good intentions dissolve in a sea of disagreement and lack of cooperation. In the end, we're left with the feeling that meetings are a waste of time, another way to procrastinate, or a way to avoid making a decision. (In fact, many times we've seen a project assigned to a task force as a means of ensuring its death!) What's even worse, the individuals participating in this type of meeting feel so frustrated that they become convinced the problem is unsolvable. They may tune out at future meetings or find excuses not to get involved.

As our example illustrates, poorly run meetings give participation a bad name. Yet, without participation an organization will lose the battle to improve customer satisfaction. The challenge, then, is to find an approach that fosters creativity, consensus, and commitment while reducing the conflicts and shortcomings common to group problem solving.

Fortunately, the techniques for improving meeting effectiveness are part of the branch of social science known as organizational development. Before trying to use them, however, we must understand why a meeting is called and look at what happens during it. We believe there are three reasons for calling a meeting: information, analysis, and problem solution. These reasons clearly affect what happens during the meeting—the elements of *process* and *content*.

WHY MEET?

Information, as the name implies, involves a clear exchange of facts. The reason for this type of meeting is straightforward and practical. For example, it may be a monthly get-together in which members of a unit report on progress toward targets. It may be a meeting called to introduce employees to a new or revised policy or procedure. Although some "selling" of ideas may be involved, the questions and answers that come up are intended to further the participants' understanding, rather than issue an invitation to open debate.

Analysis implies that participants make a more active contribution. The discussion draws on the expertise of group members to explain why something is happening. The emphasis is on problem identification, getting beyond symptoms to basic causes. Often, this may be all an individual member needs to get him or her on the right track to solving the problem. Here, for example, if a unit is experiencing problems in meeting goals, its members may come together to identify contributing factors from their varying perspectives.

Problem solution, on the other hand, is a greater challenge for a group. It goes beyond analysis to identify and evaluate solutions. Members of a problem-solving group share responsibility for selecting, agreeing on, and implementing a solution. This type of meeting brings about change.

No meeting has purely one purpose. Every meeting combines elements of information, analysis, and problem solving. The question is one of degree. On the one hand, a strongly informational meeting is common and generally easy to handle, because the decisions are already made. However, the meeting whose main purpose is problem-solving presents a challenge for even a skilled manager or consultant. This is because open communication leads to conflicts. Before any progress can be made, the conflicts must be resolved.

Conflicts often have little to do with the goal of the group. For example, there are very few people who want to provide poor service to customers; we all want to do a good job. The disagreement comes over the means, that is, the way to reach our agreed on goal. After all, it is a part of human nature to feel that our own way is the best way of doing something!

When we fail to identify or honestly acknowledge the purpose of our meeting up front, the result is often an ineffective meeting. We may tell participants we want them to solve a problem when, in reality, upper management has already decided on a solution. We may say we are looking for new ideas when, like Joan Jansen, we've already chosen the idea we want to use. Thus, we close our minds to the weaknesses of our favorite idea and to the benefits of other alternatives.

So the first rule of meeting effectiveness is understanding why you are calling the meeting and the purpose you hope to accomplish: Is the task before you one of information exchange? Are you trying to sell a decision that's already been made? Are you trying to instruct others? If the answer to any of these questions is *yes,* then the problem-solving meeting techniques described next are inappropriate.

On the other hand, do you have any of the following reasons for calling a group together?

1. You're facing a complex problem, requiring more technical expertise or knowledge of details than you have as an individual.

2. You're facing a challenge so new or different that no one can be considered an expert.

3. You clearly see several alternative courses of action and no one of them seems obviously superior to the others.

4. You've tried several alternatives already, and they haven't worked as well as you'd like. You need a shot of new ideas.

5. Whatever course of action decided on is going to require cooperation, buy-in, and commitment from several different areas if it's to be successful.

If any of these reasons apply, then your meeting can be improved by applying the group problem-solving techniques described next.

WHAT HAPPENS DURING A MEETING

Whatever the purpose of the meeting, it has two different dimensions. The first is content related. It focuses on the heart of why the meeting is called: the information to be conveyed, the problem to be addressed, the ideas generated by the group, the course of action to be developed.

The other dimension is process related: how the meeting is run, how well people interact, and how well they work with one another. This includes elements as simple as following an agenda and as complex as resolving conflict between two or more members.

In a purely informational meeting, controlling both process and content is fairly easy. The person calling the meeting generally has a clear agenda to follow—a policy or procedure to describe, findings to report on. Participants also have clear roles to follow—absorbing new information or offering information of their own.

The dimensions begin to conflict as the purpose of the meeting changes its focus to analysis and problem solution. At this point, the content becomes more complex and the process of meeting effectiveness becomes more challenging.

Consider a typical problem-solving meeting. You may be faced with a problem that must be resolved or, as in Joan's case described earlier in the chapter, you may see an opportunity for improving what's already being done. So you call together the people you think will best help you to work on your challenge.

You may tell them in detail why you're meeting before you get together or you may wait until you have the members assembled. Of course, you have to arrange meeting details beforehand: a place to meet and a time agreeable to everybody.

You want to make the best possible use of time, so you want the meeting to run smoothly. You don't want to spend too much time on unrelated issues, yet you want everyone to be heard.

As the meeting progresses, you find the ideas of others will spark new ideas of your own. In fact, so many ideas are coming up, they're hard to keep track of. At the same time, no one idea seems to offer the perfect solution and people are becoming very heated as they defend their pet solutions. Before you know it, your meeting deteriorates into a battle of egos. In the worst case scenario, group members break into opposing factions, and your meeting ends abruptly, like Joan's, with no action decided on.

Why does this happen? There are three basic reasons. First, the task of guiding both the content and the process of a meeting can be too demanding for one person to handle. Concentrating on how the meeting is going can distract one's attention from the ideas that are being presented. On the other hand, reacting only to the ideas can lead to anarchy. Ideally, both content and process must be guided and balanced.

For this reason, the first set of "rules of the game" for effective group problem solving focuses on roles. This involves a clear definition of responsibilities within the meeting. Roles are characterized as either process focused or content focused. Often, they mean a suspension of the normal organizational hierarchy for the duration of the meeting. Everyone's ideas are valued equally, at least initially.

The second reason for failure in problem-solving meetings has to do with strategy. Without a guiding structure or a systematic course of action that everyone agrees to, the meeting goes in too many directions. As a result, it arrives nowhere. On the other hand, if the strategy is too rigid, creativity may be blocked and participation dampened.

Thus, the second set of "rules of game" focuses on a clear set of steps that all members know from the outset and buy into. It ensures members that they will have a chance to be heard and that their contributions will be respected. At the same time, following a sequence of steps also guarantees that all members will have something to show for their time and effort, for it leads to clearly identified actions.

No matter how clear the roles or how logical the steps, however, meetings can still fail if the members are unable to work together in a positive and supportive manner. If members of the group don't really listen to each other, if they're insensitive to each other's feelings, if each has a different understanding of what they're agreeing to, if hidden agendas abound, the meeting is doomed to disaster. All the techniques in the world will not overcome bad faith.

So the last set of "rules of the game" deals with interpersonal communications skills. The techniques for improving group interactions are simple in theory. They often fail in practice, however, because they are process related. They become lost in the heat of the moment as the attention of group members is focused on content—the problem or opportunity that is being addressed. So our group problem-solving model not only identifies techniques but also concentrates on specific responsibilities for clear communications within the roles that group members play.

THE RULES OF THE GAME: ROLES

No matter how democratic the setting, not all members of a group are created equal. There is always a leader. This individual serves as the driving force and is responsible for seeing that agreements are developed and implemented. He or she can go by several names: the "chair" of a task force, the "project manager" of a special study team, the "boss" of the unit.

As we indicated earlier, every meeting has two elements: content and process. You can think of content as "what" a meeting is about and process as "how" the meeting is conducted. A leader must be concerned with both, aiming for the best possible solution while maximizing the contributions of each member of the group. However, as we've noted before, the task of guiding both the content and the process of a meeting is too demanding for one person to handle. Either the meeting will come to an unproductive end (the "what" is not satisfied) or the group members will leave unsatisfied (the "how" is not served.)

This is especially true for task forces and study teams with cross-functional membership. Here, the leader often contends with divergent points of view and conflicting interests, without the organizational authority to ensure that solutions will be reached and implemented. In our experience, leaders of groups usually err in favor of content. This means that the conduct of a meeting is not given equal importance. Thus begins the many problems we have discussed.

To counteract these problems, we recommend dividing leadership responsibilities in a problem-solving group between two individuals with distinctly different roles: *client* and *facilitator* (Figure 6.1). The client focuses on content, while the facilitator attends to the process of a meeting. Let's look at each of these roles in detail to see what they entail.

The *client* is the person who has ultimate responsibility for the task at hand. The task might be a challenge to install a new system, an opportunity to develop a new service, or a recurring problem that cries out to be solved. Thus, the client "owns" the task. He or she is responsible for bringing it before the group and enlisting its aid.

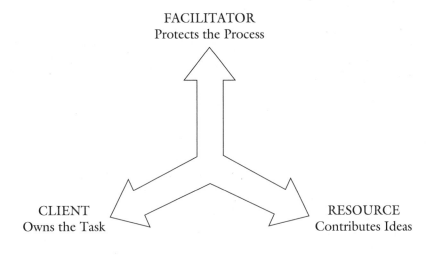

Figure 6.1: Effective Meeting Roles

What precisely does the client do during a problem-solving session? First, the client shares background information. This cues group members into how the situation came about and what has or has not been tried. It's also an opportunity for clarifying any terms or jargon involved. This is the time when the facts of a situation are presented. Data that have been gathered and analyzed may also be presented.

After setting the stage with background information, the client develops a problem statement. This statement is really a call to action. A problem statement spells out what the client would like the group to do in clear and precise terms.

As the group works on the task and begins to develop alternatives, the client shares responsibility for the content of the meeting. He or she is free to join in and offer suggestions. However, selecting and prioritizing alternatives are strictly the responsibility of the client, who reviews the ideas offered in the meeting and decides which warrant further evaluation. This does not mean that any ideas are discarded. Instead, the client balances the need to ensure that all ideas are given a fair hearing with the constraints of time and other resources.

The final major responsibility of the client is to ensure that action takes place. This means that he or she accepts ultimate responsibility for completing its "next steps," following up on assignments, and providing resources or assistance, if necessary.

While the client owns the task and concentrates on the content of the meeting, the *facilitator* focuses on how well the meeting is progressing. While the client "owns the task," the facilitator "guards the process."

The facilitator has a caretaker role, which demands a neutral stance in relation to the task at hand. This means that a facilitator's participation in the meeting is limited to directing the process, rather than contributing ideas, evaluating suggestions, or developing solutions. To put it bluntly, a facilitator does not compete with the other members of the group on content.

A facilitator has one major responsibility: to make sure group members follow the "rules of the game." This means having all members of the group operate within the bounds of their assigned roles, ensuring that the steps of the problem-solving process are followed and fostering effective interaction.

How does a facilitator do all of this? Principally by modeling the types of behaviors that increase group effectiveness. For example, a facilitator can work with the client before the meeting to prepare background information and formulate the problem statement. During the meeting, the facilitator uses a flip chart to publicly record the key points, ideas, and agreements made at each step in the process. In this way, the facilitator helps to safeguard ideas, preventing their early rejection. The facilitator also protects members of the group from attack through good communication skills and encourages the active participation of all members. By keeping the group on track and on time, the facilitator aids in making use of a scarce resource—time.

Dividing the leadership role between client and facilitator brings us to the role played by everyone else in the group—the role of resource. Resources are contributors of ideas, which means they concentrate on content and focus on the task at hand. They generate ideas and build on each other's suggestions. Instead of shooting down an idea with a cry of "that'll never work," they try hard to overcome any shortcomings in the idea and figure out a way to make it work.

Resources are content oriented. Focusing on content, however, does not mean that they judge the value of the client's task or why it's being worked on. Their aim is simply to address it. Likewise, focusing mainly on content does not absolve the resources or client of responsibility for supporting the facilitator's efforts to enhance the effectiveness of the process. All members of the group should practice the same communication skills modeled by the facilitator.

THE RULES OF THE GAME: STEPS

No matter how clear-cut the roles, problem-solving meetings will not achieve any lasting accomplishments without a systematic course of action—or series of steps— to which everyone agrees. The steps listed next are not extraordinary; they are common to any problem-solving situation. When used in tandem with the roles already described, however, they provide a strategy for group problem solving that is both logical and flexible.

A clearly defined set of steps encourages group effectiveness in two ways. First, it provides all members with a level playing field for contributing their ideas. It guarantees that ideas will be both heard and protected. Second, it provides an excellent motivation for participative problem solving—a feeling of accomplishment. Group members see their time and effort translated into clearly identified goals, actions, and results.

Our problem-solving strategy contains seven steps (Table 6.1).

Table 6.1 Seven-Step Strategy for Problem Solving

1. Sharing background information
2. Developing a problem statement
3. Brainstorming
4. Selecting and prioritizing alternatives
5. Identifying "benefits" and "concerns"
6. Overcoming "concerns"
7. Developing an action plan

Sharing Background Information. To solve a problem or to respond to an opportunity means we need a clear understanding of what we are going to be called on to do. Sharing background information lays the groundwork for a group's future efforts. Ideally, this step clarifies, to everyone's satisfaction, why the group has been convened and its mission. The time spent on this step also serves as an opportunity to evaluate whether all the right people are involved in the problem-solving effort, because the sharing process soon makes it clear if the group needs additional resources.

But who exactly participates in this part of the process? Because background information defines the task at hand, the client is responsible for providing it. This does not necessarily mean the client knows all relevant information or spends the first half of the meeting lecturing group members. He or she may call on others with more technical knowledge or experience to present the background information. In fact, some members of the group are often selected to be resources because of their special perspectives on the subject. Other times, background information in the form of a memorandum or white paper is sent to group members before the meeting. This helps group members come to the meeting already informed and ready to ask questions. Any special studies, data or information that illuminate what is really happening should be shared during this step.

The key is preparation, which is why we recommend that the client meet with the facilitator before the meeting. The facilitator can ensure a smooth start to the process by asking the client the questions listed in Table 6.2.

Table 6.2 Premeeting Questions for the Client Role

---

1. Do any key terms need to be defined?
2. Why is this a problem, opportunity, or challenge? Why is it a problem, opportunity, or challenge for *you*?
3. What have you already thought of or tried?
4. Has everyone who could give related information on the problem been questioned?
5. What would you most like to get from this meeting?

When inviting people to be members of a problem-solving team, it's important to let them know as much in advance why they are being asked to meet. This information helps them adjust their mental gears and organize their thoughts. It is an opportunity that the training task force members in our story didn't have! It's often helpful to summarize some of the answers to the questions just listed in the invitation to participants. As mentioned earlier, prereading is helpful, especially for technical or complex subjects.

How long should the sharing of background information take during the actual meeting? As long as it takes for the group to develop a good understanding of the problem. This implies, however, the need for good communication between members—questioning, clarifying, and paraphrasing what has been said, recording the key points made. The facilitator assists this process, asking questions if group members seem reluctant and making sure pertinent information is recorded in a prominent place for all members to see.

The facilitator also protects the client's problem from criticism, reminding members that it belongs to the client and should not be evaluated. On the other hand, a client may gain a new perspective on the problem by sharing background information. The discussion during this first step frequently leads the client to see that what originally appeared to be a problem is merely the symptom of a deeper or more fundamental issue. Focusing on the root cause of a problem rather than a superficial symptom saves time and effort.

Furthermore, during the sharing of background information, gaps in knowledge become apparent that lead to additional research and analysis. In fact, the sharing of background information about complex tasks can require several meetings. But only with a common understanding of the task can the team begin to address it. That's why we emphasize devoting sufficient time and energy to sharing background information before moving on to the second step—the development of the problem statement.

Developing a Problem Statement. Problem statements are developed for very simple reasons: to focus the group's energy on the task at hand and to provide it with a clear sense of where it's going. A good problem statement motivates people and makes them want to take action.

Developing this type of problem statement is relatively easy if we've done a good job sharing background information. All we need to do is express our objective as a concise statement with the following elements (Figure 6.2):

"How to"

+

an action verb

+

answers to the questions
"how much?" "how soon?"

Figure 6.2: Formulating the Problem Statement

Why do we insist on such a formula? First, the focus on "how to" accomplish a specific objective urges the group to be action oriented, rather than passive, uninvolved, or too ready to accept the status quo. Second, the quantifications of "how much" and "how soon" help the group to see the parameters by which its success can be judged. For example, consider the contrast between the following problem statements: "Too many forms are sent to the wrong sales offices" compared with "How to reduce the number of forms sent to the wrong sales office by 50 percent within the next six months."

Who develops the problem statement? Once again, the client as owner of the task has prime responsibility for this step in the process. In a prework meeting, the facilitator often helps the client develop a preliminary draft of the problem statement. However, as we've already indicated, the sharing of background information may disclose some gaps in the client's understanding of the problem. As a result, he or she may modify the problem statement or develop a completely reworded one during the meeting. Flexibility is the key: Clients and facilitators need to recognize that the problem and the process for addressing it are not written in stone.

Before moving to the next step in the process, the final problem statement should be clearly written and prominently posted in the meeting room. This chore is the responsibility of the facilitator, who either publicly records group responses personally or asks another person to handle the task. The important point is to have the statement available for easy reference by all group members. This maintains their focus on the task as the problem-solving process opens up to its more free-wheeling and creative part: brainstorming.

Brainstorming. Much of our thinking is bounded by imposed logic and rules. We search for the "correct way" to do something, the "right answer" to a question, the "correct solution" to a problem. As we grow and develop, logic and rules help us to cope with the challenges of daily life and the massive amounts of data before us. They set boundaries and help us to pigeonhole facts and classify ideas.

At the same time, however, these boundaries limit us, for they give birth to a "killer instinct" not directed toward people but toward ideas. It grows out of our sense of logic and makes it difficult for us to accept, or even see, a new idea that is contrary to our existing set of rules. This instinct leads us to evaluate an idea or alternative against the prevailing rules, to find it wanting, and to dismiss it out of hand. It thus serves as a killer of creativity and innovation.

Brainstorming is where we begin to develop alternatives. Both client and resources concentrate on the task, generating as many ideas as possible. The facilitator guards the process and ensures that the ideas are recorded on flip charts. This prevents the loss of ideas and provides the seed for developing additional ideas. At this time, the group needs to call on all its creative power. While the killer instinct can strike at any stage in the problem-solving process, it is no wonder that it makes itself felt most strongly during brainstorming. It reveals itself in knee-jerk reactions destined to kill ideas. For example, group members say: "That will never work," "It costs too much money," or "The boss will never buy that."

The facilitator does much more than record ideas during the brainstorming step. He or she also protects and nurtures ideas by enforcing certain principles. First, the facilitator tries to eliminate censorship. Ideas are often censored not merely by the group but by the individuals who offer them. A sure sign of self-censorship are statements that begin, "This sounds stupid, but..." or "I know this will never work, but...." A facilitator encourages members to express all ideas, no matter how farfetched they seem or how much their originators might doubt them.

Likewise, the facilitator does not allow other members of the group to censor or judge ideas. This derives from the second principle of brainstorming: separate producing ideas from evaluating them. This does not mean the ideas will not be judged; it merely postpones the evaluation. When we react to an idea negatively and point out all its weaknesses, we dampen the group's creativity. More seriously, a negative orientation has a divisive effect on the group. If one member is quick to point out the flaws in another's idea, the other will be looking for a chance to retaliate. Others will mentally withdraw from the discussion, seeing criticism of their ideas as criticism of themselves.

The third principle a facilitator enforces is to get the group to go for quantity and judge the feasibility of ideas later. The aim here is to get as many ideas as possible up on the flip charts for all to see. After the first burst of ideas occurs, a facilitator can often trigger more by repeating the ideas that have already been offered.

Recapping ideas also supports the fourth principle of brainstorming: building on each other's ideas. One of the chief reasons for working as a group is to benefit from the different members' perspectives. A chance remark by one person can set off a whole new train of thought for someone else. The facilitator encourages members to embellish, enhance, and expand on each other's ideas.

Brainstorming is a challenge for a facilitator; not only does it call for the skills of a referee to get client and resources to follow the rules but it also requires self-policing. A facilitator must remain neutral and concentrate on process. However, it is easy for the facilitator to be drawn into the content of the meeting, to want to offer his or her own ideas or build on those of others, particularly if he or she has expertise on the meeting's content. A facilitator must guard against this tendency. The ideas publicly recorded on the flip chart are those that clients and resources have offered and do not reflect the facilitator's personal bias or viewpoint.

Selecting Alternatives. Effective brainstorming yields a wealth of new ideas and alternative approaches. How do we decide which one best addresses our problem statement? At this point, a decision has to be made to begin the evaluation process and to determine the order in which ideas will be judged. This decision is the responsibility of the client, the owner of the task.

Frequently, the facilitator calls a break at this step and confers privately with the client. The client may decide that certain ideas fall into logical groupings and may wish to rearrange them that way, or the client may want to begin with ideas that represent a totally new approach to the problem or ideas that can be implemented in a short time. Whatever the reason, the client selects the ideas that should be evaluated. This does not mean that ideas are discarded out of hand; it merely establishes the order in which they'll be examined. Also, the facilitator encourages the client to share with the group the criteria he or she used to select or set the priority of ideas.

Identifying "Benefits" and "Concerns." How do we judge ideas? We begin by recognizing that there is no perfect solution to any problem, no perfect way to address any opportunity. Instead, there are a number of less-than-perfect alternatives. Each alternative has its strengths—benefits that can be readily identified. Likewise, each has weaknesses or gaps that can cause concern.

In this step, we look in turn at each of the ideas selected by the client. We begin by identifying all of the benefits.

1. How much does the idea contribute to solving the problem?

2. How much will it save us in resources—work-force, money, materials, and time?

3. How easy will it be to implement?

4. How acceptable is it to all parties involved in its solution?

Both client and resources identify the benefits, while the facilitator sees that each is captured on a flip chart.

Once benefits are recorded, the group can turn to areas of concern. Again, the client and resources identify the concerns and the facilitator sees that they are recorded. Concerns are the flip side of benefits.

1. What parts of the problem can't be solved by this idea? Can we live with them?

2. Do we have the resources—workforce, money, materials, and time—to implement the idea?

3. How difficult will it be to implement?

4. Can we expect resistance from key people or from any of the parties we need to involve in the solution?

Just as we follow certain principles in the brainstorming stage, we follow two guidelines as we analyze the strength ("benefits") and weaknesses ("concerns") of each idea. First, benefits, strengths, and advantages must be identified before the group identifies the idea's weaknesses, concerns, or disadvantages. A good facilitator, in fact, will insist on at least three benefits before accepting any concerns. Once again, we're trying to counteract the "killer instinct," to nourish an idea and give it a chance to grow.

Our second guideline is to phrase concerns as action statements, beginning with the words "how to." As we noted in our discussion on developing problem statements, this format encourages action and discourages passive acceptance of the status quo. It then becomes easier to resolve concerns by recycling them as problem statements. It thus sets the stage for the next step in the process: overcoming concerns.

Overcoming "Concerns." No idea is absolutely perfect; even the best of ideas has weaknesses or gaps. In some cases, the gaps are insignificant and are more than counterbalanced by the benefits. In other cases, we may be able to transform the idea into an acceptable solution if we can figure out how to fill or minimize the gaps.

In this step, the client plays a decisive role, weighing the concerns that have been identified. With the help of the group's resources, he or she determines which concerns can be lived with and which render the idea unusable as it now stands. The concerns that render the ideas unworkable must be overcome. The group does this by recycling them through the problem-solving process.

The concerns become new problem statements—after all, they're already phrased as "how to's." We then repeat the same steps: brainstorming ideas to solve the problem, identifying benefits and concerns, and overcoming concerns where necessary, until we we arrive at a solution that will help us modify our original alternative, transforming it into an acceptable idea.

It is important to realize that not all gaps have to be filled or disadvantages addressed fully for the solution to become workable. The challenge is to determine which gaps, or "concerns," need to be overcome and to see if a plan can be developed to do so. For some ideas, it may be possible to ignore the concerns altogether; for others, the cost of overcoming concerns may more than outweigh the benefits. All of this is taken into consideration as the client takes the lead in developing "next steps," the final stage of the problem-solving process.

Developing "Next Steps." This is where words meet deeds. This step identifies who is going to do what and when it is to be done. In other words, developing "next steps" is designing an action plan. This gives members something to show for their contribution of time, effort, and knowledge, as clearly defined assignments increase the probability that action will take place.

While the client takes the lead in developing the action plan, resources contribute to this stage of the process. Together, they work on identifying next activities, assigning responsibilities, planning the sequence of activities, and putting into place control mechanisms that will monitor progress and indicate results.

The facilitator ensures that all parts of the action plan are clearly recorded for client and resources to see and agree to. This step, more than any other, distinguishes problem-solving groups from the typical task force or committee. It makes planning and feedback an integral part of the meeting—and it gets these aspects into writing before the members disband. After the meeting, the facilitator makes sure that what has been recorded during the meeting is typed and distributed in a timely manner.

Before we close our discussion of the problem-solving sequence, we wish to share a few words of warning. While the problem-solving steps are presented as neat and distinct stages in the overall process, in practice they are fluid and blurred. Group members may feel they have completed one stage, only to find themselves backtracking a few minutes later. We may be generating ideas in the brainstorming phrase then suddenly realize we need additional background. We may be listing concerns when three new benefits appear or a totally new idea comes to mind. We may also be bound by time constraints. As a result, we may have to stop our meeting after completing one or two steps.

How do we prevent such situations from deteriorating into anarchy? The answer is to remain flexible. A group may at any time return to a previous step to complete unfinished business. Likewise, whenever we have to interrupt the process and postpone what we are doing until a later meeting, we can skip ahead to "next steps." In such cases, the next step could be to reconvene. At times like these, we fully appreciate the role of facilitator to guard the process and keep us on track!

THE RULES OF THE GAME: COMMUNICATION

Defined roles and a sequence of steps provide the machinery for improved group problem solving. Like any machine, however, they still require a lubricant if they are to function effectively. Communication tools and techniques that improve group interactions meet this need.

It is easy to see how meetings fail if the members are unable to work together in a positive and supportive manner. But even groups working in good faith suffer from communications failures. Their good intentions fall apart because the techniques for improving group interactions are process related and people are used to focusing on content—the problem or opportunity that is being addressed. That is why our group problem-solving model not only identifies techniques but also 87 concentrates specific responsibilities for clear communications on the role of the facilitator. All group members bear responsibility for their interactions. However, the facilitator serves as a role model for client and resources and encourages them to use good communications techniques.

The six basic techniques for improving group interactions are paraphrasing, reflecting, open-ended questioning, headlining, reacting with benefits before concerns, and using flip charts. Very often, conflict arises in a group situation because of misunderstandings. So the first four techniques—paraphrasing, reflecting, using open-ended questions, and headlining—are designed to enhance communications between group members. The next technique—"reacting with benefits before concerns"—has been discussed as a step in the problem-solving process. However, it's applicable to other parts of the meeting as well. The last technique—using flip charts—visually reinforces group understanding and becomes the group's memory.

In the following section, we discuss each technique in light of our story, "If You Don't Want to Listen, Why Ask." The examples highlight each technique and show how they're used to improve group problem solving.

Paraphrasing

When we paraphrase, we restate, in our own words, what someone else has just said to make sure we have understood. This is done to prevent misunderstanding? We signal we are doing this with phrases such as "It sounds like…," "What I hear you saying…," "It seems like…."

For example, in the training team's meeting, Alicia said: "We're already behind schedule, and I'm worried that the extension we were granted will not be enough—and that's without major modifications to the design."

The facilitator or another member of the team could paraphrase Alicia by saying: "Alicia, it sounds like you think we may not meet our target even if we don't change the design."

Paraphrasing both clarifies and summarizes what is being said. As such, it can be used at any time in the process, from the sharing of background information right on up to "next steps." And while the facilitator uses it to ensure that individual contributions are correctly recorded, all other members should paraphrase as well to increase individual understanding. Paraphrasing also improves group participation. It sends a signal to speakers that their ideas are valued and listened to.

Reflecting

Just as we use paraphrasing to ensure that we understand what group members are saying, we use reflecting to make sure that we understand the feelings that underlie their words. We mirror to the speaker the emotion or feeling that is being conveyed to us as listeners.

For example, Alicia questioned Joan: "What exactly does 'incorporating this material into our design' mean to you? If you're thinking of any changes in the lesson plans, count me out!"

To reflect on what Alicia is saying, a facilitator, Joan, or any other member of the group could respond: "Alicia, I can see that you're upset by the additional work this will add to the training program."

Reflecting demonstrates our desire to connect to the speaker and how he or she is affected by an idea or suggestion. As with paraphrasing, it shows individuals that their contributions are valued and their feelings are respected. In these ways, it can enhance participation.

Open-Ended Questioning

When we phrase questions in such a way that they require more than a one-word or a *yes* or *no* answer, we are using open-ended questions. These questions allow a person to open up and tell us more. We ask them to get additional information, to clarify what has been said, and to obtain feedback.

Open-ended questions begin with words like "what," "why," and "how." For example, in the training meeting, Marla said: "…what about giving the book to the participants before the seminar, as prereading?"

Instead of attacking the idea as naive, Roger could have acknowledged her idea and then used an open-ended question: "That's a good alternative we should consider. How do you think we can get the participants to read the material before they come to the training?" Marla would have been able to contribute any additional thoughts on how to apply her idea without feeling she was under attack.

This last example also points out another benefit of open-ended questions: drawing shy or retiring group members into the discussion in a nonthreatening way.

Headlining

This technique allows us to capsulize what is being said before we expand on it. Just as the headline in a newspaper signals what is in the story that follows it, headlining during a meeting signals to our listeners where we're headed and the point we wish to make. As an added benefit, these succinct descriptions also simplify the work of recording what takes place in the meeting.

In normal conversation, when offering our ideas, we spend a lot of time leading up to the main point. Listeners sometimes miss the important part. For example, at the beginning of the training meeting, Joan offered several related ideas:

"I'd first like to point out what I see as the positive effects this addition would make to our program. The book deals with complex ideas in a straightforward manner, something much appreciated by today's managers…. By incorporating this new material into our current design, the managers—our customers—will benefit directly."

As the last part of the story indicates, her main point was using the material to make the training program even better. A headline such as "The material in this book can improve our training program" would have enabled her listeners to see her main purpose. It also would have described an end result with which few would argue.

Headlining can be used at any stage in the process—from background information through "next steps." A good facilitator will set the example, offering to "headline" what a speaker is saying and using the words of the headline on the flip chart. As participants become more familiar with the technique, the facilitator can then ask group members to headline what someone else has said or their own thoughts before expanding on them.

Reacting With Benefits Before Concerns

We've discussed identifying benefits and concerns as a separate step within the problem-solving sequence and pointed out that benefits should always precede concerns? However, concerns can and do surface during any part of the process. That's why it's important to recognize and counteract our natural "killer instinct." We can do it simply by resisting the urge to attack an idea with an acknowledgment of its good points: "Good idea, I've never looked at the issue that way," or "This has sparked another thought. I'd like to build on your idea."

What do we do with our concerns? We save them until the appropriate time in the process we are called on to deal with them—the steps during which we identify benefits and concerns and try to overcome the more significant weaknesses. Until that point, we can capture our personal concerns with an idea on a notepad, phrased as "how to" statements.

Using Flip Charts

Throughout this chapter, we seem to "flip out" over the use of flip charts. At every step along the way, we affirmed the facilitator's responsibility to ensure that what takes place is recorded on flip charts for all members to see rather than rely on a secretary taking notes.

Flip charts are used for many reasons. First, they focus the group's attention. They are particularly helpful for groups who have little experience with the problem-solving roles and steps, for they help group members to identify how far along they are in the process. Flip charts also offer immediate confirmation that all members of the group share the same understanding of what is happening. Speakers see immediately if their ideas have been understood by looking at what appears on the chart. And, by using two senses—hearing and vision—we improve reception and understanding of ideas.

Use of flip charts also helps to ensure that no ideas are lost. Group members can see if background information, ideas, benefits, or concerns have been omitted and make sure they're added. Finally, writing "next steps" on flip charts before the meeting concludes increases the possibility that the appropriate actions will take place after the meeting. Any misunderstandings concerning assignments or target dates can be cleared up while all the parties involved are still together.

SUMMARY

In this chapter, we've explored the benefits of group problem solving. We've also looked at some of the reasons why groups fail to achieve these benefits, such as conflict between the content and process of a meeting, lack of a clear common strategy for analyzing the problem, poor interaction between group members, lack of basic ground rules, and organizational conflicts and hidden agendas.

To counteract these weaknesses and to ensure employee involvement, we recommend a process that contains three elements: responsibility for content and process divided between clearly defined roles, a sequence of problem-solving steps, and interpersonal communications techniques designed to enhance understanding and participation.

Armed with the problem-solving process, individual participants in a customer-provider work flow can work more effectively as a team to improve customer satisfaction.

QUESTIONS FOR DISCUSSION

"If You Don't Want to Listen, Why Ask?"

1. Who should have served as client in this meeting? What was the issue the client wanted to address? Was it a problem or an opportunity? Try phrasing it in the form of a problem statement.

2. How many ideas did the team come up with? Identify each and then indicate the benefits and concerns that team members identified for each.

3. How did members kill ideas? Give specific examples.

4. What points did everyone agree on?

5. Did all members participate equally? Why or why not?

6. How do you think each team member felt furing the meeting? How did each feel after it?

7. What techniques could have been used to improve the meeting? Give examples of where they could have been applied.

8. Consider one of the concerns the group came up with (which you identified in question 4, above). Be sure it is phrased as a "how to" statement and cycle it through the problem-solving process.

HOW MANAGEMENT PRACTICES AFFECT SERVICE QUALITY

In Chapter 4, we introduced the idea of using the customer's viewpoint to determine what service quality is for your organization. In Chapter 5, we discussed a way of measuring customer feedback that gave us quantifiable information that pinpoints the areas that must be addressed if we are to improve service quality. In Chapter 6, we discussed why it is critical to include all employees in the process of improving customer service. Without their active involvement, improvement efforts are doomed.

These ideas signal a critical shift from the way traditional organizations view their relationship to customers. They also lead to certain conclusions about an organization's success that have important implications for service providers.

Conclusion 1: Improved service quality enhances profitability by stemming customer defections. Services retain customers by meeting and, where possible, exceeding customer expectations. Retaining customers is the key to an organization's growth and survival service, because it is much less costly to retain existing customers than to acquire new ones. In other words, long-term customers are generally a business' most profitable ones.

Why is this so? The process of acquiring new customers is a costly one. At a minimum, it involves marketing and sales costs. In many organizations, the cost of establishing new accounts—contract development, credit checks, and the like—can also be significant. However, only recently has any attention been paid to the cost of losing an existing customer. This cost represents the lost income and profit that a satisfied customer generates over time.

Professor W. Earl Sasser believes that today's accounting systems don't capture the value of loyal customers.* In an article he wrote with Frederick Reichheld, they describe a customer's real worth as the sum of the projected profits over the life of the customer-provider relationship. As a company reduces customer defections, the average relationship lasts longer and produces larger profits. Thus, high-quality service makes good economic sense. Not only is it the key to attracting new customers, it is also a means of retaining existing ones.

Conclusion 2: Personal referrals are a key way that service providers attract new customers. No matter how complex or sophisticated a service, it ultimately boils down to one customer at a time interacting with one member of a service organization.

Unlike manufactured goods, where similar products from different producers can be matched head to head for appearance, size, ease of use, and other tangible performance measures, service performance is a direct reflection of the satisfaction of existing customers and the personal experience they had with the service company's employees. Therefore, one of the major factors in getting new customers is the word-of-mouth recommendation of current customers. Those endorsements will come only if current customers are convinced that the service they receive is, at the very least, meeting their expectations.

Conclusion 3: While the more easily measured outcomes of a service—the aspects that make up the reliability dimension—must be present for a service organization to compete, it is the process dimensions that provide a competitive advantage. This lesson is not as obvious as the others. It bears certain similarities to theories of human motivation. Frederick Herzberg looked at the factors that motivate people in the workplace and came to the conclusion that there are two types: dissatisfiers and motivators.

Dissatisfiers, which are also referred to as "hygiene factors," represent those items in the work environment that form the minimum acceptable work conditions: wages and benefits that reflect market rates, a secure working environment, and the like. They define the context in which work is performed. If you do not provide these maintenance factors, your employees will be dissatisfied, and you will not be able to attract and retain a good work force.

However, adding more of the hygiene factors doesn't mean your employees will become more satisfied or more motivated to work. Instead, you must turn to the items that Herzberg called motivators that are intrinsic to the job: the challenge it provides and the opportunity to feel a sense of achievement and to advance. They define the content of work and serve as an internal feedback loop: The more people get, the more they want.

This same analogy can be applied to customer perceptions of services. The outcome dimension of service quality—reliability—is the hygiene factor. Delivery of the basic service you promise will meet the minimum standards set by the marketplace.

---

*"Zero Defections: Quality Comes to Services," Frederick F. Reichheld and W. Earl Sasser, Jr., *Harvard Business Review*, September–October 1990, pp. 105–109.

Failure to do so yields customer dissatisfaction, poor word of mouth, and loss of market share. On the other hand, increased reliability will not make customers any more satisfied. In fact, increased perceived reliability implies that you weren't accurate or dependable in the first place: Either you provide the promised service or you don't!

At the other end of the spectrum are the process dimensions of service quality. They focus on how the customer is treated and can be considered the satisfiers of service quality. The process dimensions do not present a "black-or-white," "right-or-wrong" type of choice. They reflect individual behaviors and thus offer continuous opportunities for improvement.

The process factors also drive up customer expectations. When you exceed customer expectations in these areas, customers not only want more but are also likely to demand more from your competitors. Thus, by paying attention to the relationship aspects of service quality, you can distinguish yourself from your competitors and are more likely to exceed the customer's expectations, increase loyalty, and ensure business success over the long term. Moreover, in our experience, the process factors (responsiveness, empathy, and assurance) are more difficult for competitors to copy, thereby providing you more advantages in the marketplace.

These conclusions provide further insight into why service quality makes a difference, why it goes hand in hand with market growth and increased profitability. However, we don't become providers of high-quality service simply by accepting these conclusions. Likewise, all the tools and techniques we have discussed thus far will not guarantee success. Their value rests in their ability to help us define what we do and to pinpoint the areas in which we need to improve. They provide a starting point for action.

To provide high-quality service successfully, we must actively identify the root causes of service quality problems and eliminate them. Otherwise, we will be in a constantly reactive mode, responding to the crisis of the moment. Being active requires a thorough understanding of the inner life of a service organization.

For the remainder of this chapter, we'll be describing a way to assess your organization's service quality maturity that provides a springboard for improvements.

ASSESSING QUALITY MATURITY IN SERVICE ORGANIZATIONS

Consistent delivery of high-quality service doesn't happen overnight. It requires changes in the way the organization operates and manages and rewards its people. In short, it needs a major cultural change. Such a transformation takes time and cannot be easily copied. Thus, a competitive advantage results from the practices we describe throughout this book.

How much change is needed, and where do we have to concentrate our efforts? We have found it helpful to have a framework that allows us to examine an organization in terms of its ability to provide its customers with high-quality service. We call this framework the "Quality Maturity Scale" (Figure 7.1.) It is a way of assessing your current situation and identifying areas for improvement. Figure 7.1

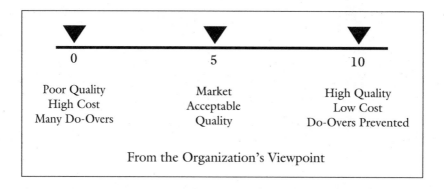

Figure 7.1: The Quality Maturity Scale

shows values from 0 to 10. To better understand how to use this scale, let's look at the organizational characteristics found at each extreme.

THE VIEW FROM INSIDE THE ORGANIZATION: SYSTEMS AND PROCESSES

At the lowest end of the scale, at 0, nothing is going right. Extensive do-overs hamper efficiency. Errors overwhelm the work process. What goes out the door is not acceptable in the marketplace, because shoddy services are being provided to customers.

The "0" organization creates little profit as a result of the high cost of doing business. Its energies are focused on dealing with complaints and trying to stop the steady stream of customers who are deserting it. As a result, few resources are available for the development of new products and services, and it makes weak responses to market changes. This lack of flexibility results in a marketplace perception that the organization is "just reacting" to change. Innovation never gets started or is so hampered by internal difficulties that it never goes anywhere.

Only fragmented pieces of the work process are documented or even understood in the "0" organization. Thus, problems identified and solved for one part upset other parts of the process. Errors are often passed on to customers. Extra-processing is the normal way of doing business, causing high overhead and operational costs.

Customer contact employees may be blocked from sharing their insights with the organization's decision makers. A heavy bureaucracy, with many management layers, intervenes between top management and service providers who are on the firing line. Communications from and to the top are unwieldy, often garbled, and, in some cases, totally (and intentionally) cut off.

In contrast, the "10" organization produces its products and services at low cost because it has streamlined its work flow. It has eliminated or greatly decreased extra-processing; costly do-overs are a distant memory. It does not rely on inspection to achieve marketplace acceptance. This organization prevents errors by building controls into the work process at various stages. Thus, errors or defects are detected early in the process, and problems are addressed before they become too complex to solve easily. The organization looks at transactions from the customer's viewpoint, which means that the work process is understood as a whole, not merely in individual steps. Work processes are well documented and under review for continuous improvement. Service quality, to the extent possible, is designed into the work-flow from the start.

THE VIEW FROM INSIDE THE ORGANIZATION: PEOPLE AND RELATIONSHIPS

In the "zero" organization atmosphere, only the people who remain "uninvolved" succeed. They fall into three major types. First, there are people in the zero organization who say, "It's not my job." Their attitude reflects a feeling that initiative will be punished and innovation discouraged. Their fears are well grounded, for the "zero" organization views any change as a threat, whether from inside or outside. It incorrectly assumes that the world is static and people don't have to think. After all, in an unchanging world, everything has already been defined.

The second type of people found in the "zero" organization believe "knowledge is power, so don't ever share it." They're not concerned about serving customers; they're busy protecting their own "turf." Jobs are very narrowly defined. The organization suffers the same fate common to large bureaucracies, public and private, that is, a lack of sharing and communication between functions or levels. No one, not even managers, seems to understand the "big picture." Discussions of mission or values are nonexistent, and goals remain vague, if communicated at all.

The third group's attitude is characterized by the phrase, "I only work here." It reflects the lack of personal responsibility common to people who are frustrated and excluded from decisions that affect their work. In fact, people other than those in customer contact jobs don't know who their customer is—or that they have a customer. They withdraw from the organization and each other to protect themselves from a system they feel has beaten them. In a manufacturing environment, this leads to accidents, poor production levels, and defects. In services, the picture is even more bleak. This is because the delivery of services contains many of the same elements as personal relationships. For services, the process (the way a provider delivers a service) is as important as the outcome (what finally goes to the customer.) In this organization, "no one's home," and the customers are left feeling they are dealing with automatons.

Employees don't care about their work because they have no input into the decisions that affect their work. First, they may have been inappropriately placed in a work situation that does not make full use of their skills. Then, they aren't provided with the tools, information, or training to do the work properly. For example, they work with a computer system that is constantly unavailable.

Customer contact employees are trapped in a web of inflexible rules that don't serve the customer and somehow manage to face that customer everyday by making   97 excuses for those rules. Employees may even be placed in the terrible conflict of having to do things that they know from past experience customers hate. For example, they are required to demand that customers fill out complicated forms in triplicate or stand in unnecessarily long lines.

People working in an organization that is failing to keep old customers and acquire new ones are anxious about the future. Turnover soars; morale declines as people see others leave. Managers in this type of organization frequently talk about cost overruns, budget problems, and pricing deficits. They believe that their employees don't work hard enough, but they themselves do little except exhort them to "try harder and do better." Typically, these managers blame their problems on circumstances that are beyond their control. They fail to raise their sights outward to the customer and thus remain focused on the short term. As leaders, they have no vision to share. They do not create trust, which is the "glue" needed to cement the relationships that are critical to producing quality service for customers.

The ideal situation is at the other end of the scale, when we look at the "10" organization. A "10" organization is flexible, adaptive, and responsive to change. Here, managers don't just talk about market research, customer needs and expectations, goals, and mission. These concepts drive their planning. Managers make decisions in a timely manner and share authority with all levels in the organization. Horizontal communication between organizational units is frequent, focused, and productive. This is possible because everyone's priority is to provide value to the customer.

People who work for this kind of organization show concern for their work. They feel respected because their ideas are sought and contributions valued. People in a "10" organization are responsive to change and anticipate opportunities. They regularly come together across organizational lines to solve problems. They feel committed to learning and personal growth, and the "10" organization supports them in their efforts.

THE VIEW FROM OUTSIDE THE ORGANIZATION

The scale described above took into account the internal workings of any service organization. So far, it described the organization from the inside out. However, there is another view to be considered: the customer's. If we superimpose the customer's viewpoint on our scale of quality maturity, it now lets us look from the outside in—the way the customer sees the organization (Figure 7.2).

In the "0" organization, as shown in Figure 7.2, service is not meeting customer expectations. Performance standards (if they exist at all) do not reflect what is important to customers. Because this organization is focused inward, managers base effectiveness measures on such things as their own convenience, what was done in the past, what they think is possible, and what *they* believe is good for the customer.

Customers are viewed as an amorphous group that is frequently the cause of problems. Complaints are commonplace. Customers are always sending work back

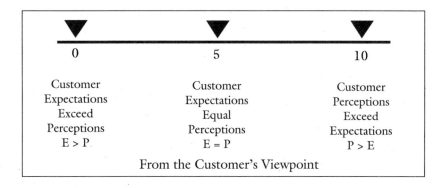

Figure 7.2: The Quality Maturity Scale

to be redone or replaced, and many of them are leaving without complaining. They are simply moving to another provider, too frustrated to continue dealing with the "zero" company, and they leave without explaining why. However, you can be sure they are telling all their friends, family, and colleagues. Customers may also be appealing to governmental agencies or consumer hotlines, in the hope of getting someone to listen to them. As a result of this lack of customer focus, sales go down and the percent of market share drops.

A frequently overlooked reason why this happens is how the organization communicates with its external customers. It inaccurately depicts what is actually happening by promising more than operations can deliver, usually through sales contacts or advertising. This means that marketing efforts are inappropriately raising customer expectations.

For example, salespeople are misinformed by marketing that customers can expect 24-hour delivery. Yet, current processes and systems operate on a 48-hour cycle. The 24-hour cycle is being developed, but is not yet functioning. Marketing, in this case, raised customer expectations prematurely based on faulty communications with operations.

The above example seems obvious, but points to a serious deficiency in how functional areas communicate with one another. Thus, communications with the customer are distorted or blocked due to deficiencies between functional areas. The marketing people would never dream of asking customer contact employees about what advertising or other communications to customers should include.

The situation is clearly different when we look at the ideal organization at the other end of the scale. Every employee has a customer and knows what that customer needs and wants. Effectiveness measures are based on key customer expectations, which are regularly reviewed. As a consequence, in a marketplace full of look-alike service providers, the "10" organization stands out from the pack by overwhelming its customers with good service. Its customers experience flexibility, and their demands are met in a timely manner. Exceptions are handled promptly and fairly, without the need for many layers of approvals. If an error is made, customers are pleased to find that it is not compounded by more errors or delays.

Instead, they are impressed at how well the organization can recover and fix what is wrong. The business advantage is clear: Customers believe they are getting value for their money and recommend the ideal organization's service to their friends and relatives.

A REALITY CHECK

Before we go on to discuss how you might use this scale to assess your own organization, we thought you'd enjoy visiting someone else's. As you read about Alice West, consider the factors we identified from inside the organization that signal quality maturity. Then, armed with your knowledge of how the hospital is treating its customers, where would you place it on the quality maturity scale?

ਝਾ ਝਾ

"Take Two Aspirin and Call Me Tomorrow"

Alice West, supervisor of admissions and central billing at Local General Hospital, had not yet taken off her coat, when the phone on her desk started ringing. Just as she reached over to answer it, Carol, her assistant, came rushing into her office.

"Don't answer it!" cried Carol. "It's the new switchboard answering system again. All the phones keep ringing even though there's no one on the line and no one can get through. It's making everybody crazy!"

Alice just sighed. The new system, LITESPEED, was supposed to solve all the hospital's phone problems, internal and external. It was so sophisticated that the consultant from the telecommunications company spent weeks on site fine-tuning it. Alice thought they would never get rid of him.

However, when Alice asked about training for her staff, especially the switchboard operators, the consultant wasn't very helpful. "Tell them to read the manuals," he told her. "But the manuals are so hard to understand. It seems you have to have an engineering degree to understand the simplest directions," Alice responded.

His answer was not comforting. "In my opinion, training is just an excuse for not using the system." Then he added, "I'm sure you'll do just fine—I understand you're all experienced people."

Alice was furious at his patronizing attitude and took her case to her boss, the director of administrative support. Her pleas for training fell on deaf ears. Instead, the director lectured her on the general benefits of strategic planning. Installing automated systems was part of the hospital's "new five-year plan." Privately, Alice wondered what this new plan was all about and if it included decreasing her staff.

"It's never the doctors or nurses who suffer when the hospital looks to cut costs. It's always the 'expendables'—switchboard operators, clerical staff, dieticians—in other words, us," she had confided to Carol.

The telephone system wasn't the only computer problem Alice faced.

A year ago, the hospital invested in a new automated billing system. Alice's boss proclaimed the end of all the difficulties associated with billing patients. After all, it was his report that recommended buying the computer equipment and software. Yet, after months of unresolved systems difficulties, her staff used the automated system only as a backup to the old manual system. The clerks simply didn't trust the new system, so they continued to process bills using the old procedure. Even though it was cumbersome, at least it was familiar.

Alice knew that her staff members were being stubborn about using the new system—they couldn't get used to using a keyboard constantly. On the other hand, she also understood why they were reluctant to rely on the information the system provided.

For example, the new system was responsible for causing severe problems for patients who had never been billed yet were being contacted by the hospital's collection agency for nonpayment. These patients were all covered by Medicare or private insurance. Yet, part of their expenses had never been included during the initial billing, and the hospital was never paid for the services rendered.

Six months later, these patients were on the phone asking why their names were in the hands of a collection agency—and her staff had no answers. The number and intensity of irate phone calls had tripled since—the year before the new system was installed.

Carol had figured out the root cause of the problem. When a patient checked into the hospital, there were generally a number of tests that had to be performed before treatment could begin. This was particularly true of surgery candidates. For every test needed, the patient was given a separate form, in triplicate, which defined the needed test. Patients who had multiple health problems, like high blood pressure and diabetes, could carry around 16 to 20 separate forms.

Carol went to Alice with her ideas. "It's simple. The possibility of misplacing or losing forms increases in direct proportion to the number of forms. It's really ridiculous that these patients, who are nervous to begin with, are being asked to deal with all those forms."

"I'm really busy now, Carol," Alice replied apologetically. "Maybe we can get together after this billing crisis is cleared up."

"But I'm talking about the billing crisis too, Alice. The billing clerks' morale has never been lower—we have to do something soon or they'll start quitting."

"OK, OK, we'll get to it next week."

But the minute she had said it, Alice felt guilty. It would take months to resolve the billing problems. After all, the hospital was being pressed to become more efficient, and in these cases, it had not been paid for services it provided.

Alice felt thankful for one thing—the questionnaires that patients filled out about how they had been treated by the hospital were given to them just before they checked out. Her staff made sure that they collected them before the patients left the hospital. Thank goodness nobody had thought of sending them follow-up questionnaires several months after their release. She knew the results would be much less favorable.

ze ze

It would be useful if you could share this story with one or more of your colleagues to compare and contrast your placements of the hospital on the quality maturity scale. Remember—there is no absolute right answer!

ASSESSING YOUR ORGANIZATION

The quality maturity scale is a useful tool for for any service organization because organizations practicing total quality exhibit identifiable characteristics. No one action or process alone makes for world-class service quality. When several characteristics come together, and when the right approaches are deployed, they enhance each other. They are more than the sum of their parts. Conversely, an organization must work on all the elements described if it is to rank at the higher end of the scale.

It is also important to realize that improving service quality—moving up the scale—is not without its price. For example, moving from "0" to "5" means that an organization has achieved market acceptance of its products and services. Although errors continue, the organization first copes with an inefficient work process by doing its best to find the errors and prevent them from reaching the customer. However, the "5" organization's cost of doing business may be even higher than a lower-ranked organization because it is inspecting everything (or most things) that go out the door. Endless rounds of checking pervade the work flow, absorbing resources that could be put to better use. So the organization's investment in quality must be twofold: investment in inspection to achieve credibility in the short term, and investment in prevention to achieve high-quality results at a low cost in the long term.

How can you use the quality maturity scale? First, look at each of the three components that anchor the scale: (1) your internal processes, procedures, and work flow; (2) the involvement of your people; and (3) your ability to focus on and satisfy the customer.

Under each component category, list your organization's characteristics—both the negative ones that drive you toward the "0" end and the positive ones that move you toward "10." Don't be limited by the characteristics listed in the above discussion, for we may have missed those that are unique to your organization. Add freely to the list at each end of the scale. Then, honestly place your organization along the scale. The results of this analysis, along with the organizational self-assessment and customer-provider work flows, provide you with the launching pad for improving the quality of your services.

SUMMARY

As you can see, the quality maturity scale helps you analyze your organization's current strengths and identify opportunities for improvement. It provides an effective
starting point for anyone who wants to build an organization dedicated to service

quality. However, converting this desire into reality requires more than just acknowledging current weaknesses. It calls for a deeper understanding of specific attitudes, practices, and management approaches ingrained in the organization's culture, which can contribute to poor quality.

In Chapter 8, we'll provide you with another opportunity to assess how well your organization copes with providing excellent customer service. We will present questions you can ask yourself that point to internal practices that influence business success by their repercussions for the customer. We'll also offer suggested solutions and strategies for decreasing your internal gaps, some of which have already been discussed in this book.

DISCOVERING THE ROOT CAUSES OF SERVICE QUALITY GAPS

In this chapter, we offer you a final opportunity to examine your organization in light of those factors that the most current research points to as critical to achieving service excellence. The questions that follow highlight problems areas that signal whether a particular "break point" exists in your organization. Following each set of questions, we offer possible solutions, ideas, and alternatives for your consideration, many of which are already included in this book.

SOME SOLUTIONS FOR CONSIDERATION

If you answered no to many of these questions, your first challenge is to create a customer orientation among managers at all levels. Even before knowledge and understanding, they must be willing to acknowledge customers as important.

The organizational self-assessment described in Chapter 1 goes a long way to foster that acknowledgment. The relatively simple act of getting people throughout

Table 8.1

Do Managers Know or Care About What Customers Think?	Yes	No
Have you identified the products and services that your organization produces?		
Do you survey your customers to get their perceptions of your products and services?		
Do you survey your customers to get expectations as well as perceptions?		
Do you survey your most important customers separately?		
Do you analyze complaints to help identify problems in your service process?		
Have you set up customer panels or groups that meet regularly to provide feedback on your service?		
Do you and your managers meet with customers regularly?		
Do you survey intermediate customers, such as distributors, agents, and dealers, as well as end-users?		
Have you established a baseline for customer satisfaction?		
Have you created easy ways for customers to contact you, for example, customer suggestion boxes or hotlines?		
Do you conduct telephone interviews with a random sample of customers regularly, for example, every week or month?		
Do you question customers in detail as to their experiences with your services?		
Are you isolated from everyday contact with customers? For example, are you on a high floor or in a different building?		
Do you or your managers meet with customer contact employees frequently, at least once a month?		
Do your customer contact employees have a formal process for sharing customer information with you, through meetings or surveys?		

an organization to ask "What are my products and services, who are my customers, and how do I know I'm living up to what my customers expect?" leads to profound changes. Two types of changes occur. First, there are changes in the content of what managers know about customers. Second, the *process* of doing the exercise itself begins to change the the manager's viewpoint, moving it from narrowly inward to outside the organization.

Other activities described in previous chapters also help. For example, when diagramming work flows, managers and analysts would do well to remember the concept of the "closed loop." This means that a work flow no longer represents a series of tasks completed inside the organization that lead to delivery of the end service to the customer. Instead, closed loop work flow diagrams acknowledge that the customer is an intimate part of the service process itself. A customer request for service initiates a series of tasks that flow through the organization, frequently crossing functional boundaries. The end result of all these activities is the delivery of service to the customer whose request started this whole chain of events. Thus, service begins and ends with the customer and all those involved along the way—adding value to the service transaction—are an integral part of the service system.

Of course, it is very important to encourage managers to gather customer information through a variety of means. The dimensions of service quality discussed in Chapter 4 are indispensable for understanding customers in a general way—what they want from services in general. But we need to know specifics about our customers. Therefore, it is essential to invest in customer feedback tools and techniques, such as SERVQUAL (see Chapter 5).

Another important way to reconnect managers with customers is to make sure that they meet with customers regularly. Following are examples that other organizations have tried:

1. Putting managers on hotlines or 800 numbers so they speak to customers directly. Do this frequently and regularly, such as once a month for one day or two half days. This should include top managers.

2. Managers should read letters from customers and take the time to answer some of them. Not all letters are complaints; some may be giving valuable information about the next product or service. Reading summaries of the letters is not enough. Managers have to get to know the human beings behind the comments summarized neatly in a management report.

3. If your service has a counter in local outlets, then managers should serve at these counters regularly. Nothing makes the customer more real and compelling than this.

4. It may be necessary to reduce the number of levels in your formal hierarchy if it removes managers so far from the everyday action that they no longer understand the business or its customers.

Managers should foster an atmosphere of partnership with employees so that employees will feel comfortable discussing customer problems and viewpoints with managers. This means managers should cultivate a manner which makes that approachable, not someone to be feared. Employees, particularly customer contact people, must have the confidence to speak freely about what customers are saying or how they are reacting.

Sometimes atmosphere is not enough. It's necessary for managers, particularly those at the top of an organization, to invent ways to meet employees, particularly those who meet with customers. This means inviting employees to informal meetings, creating employee advisory groups, and making time to meet with employees where they work. These activities break down the isolation of managers.

Once managers know what the customer expects, they should also concern themselves with managing those expectations. Part of managing customer expectations is "educating the customer." The research in service quality is clear that customers expect organizations to play fair. Thus, service providers need to explain their practices to customers clearly so as to avoid disappointment. Service providers need to teach customers more about the service and its design, features, and peculiarities. Most of all, service providers must learn to listen actively to their customers so that they can respond effectively to customer concerns.

Internal providers and customers have a special relationship that requires a similar management of expectations as with external customers. Perhaps internal providers don't need to do the same kind of education as with external customers—after all, they do work for the same organization. In that case, it helps to substitute the following questions when talking to internal customers:

1. What do you need from us?

2. What do you do with what we give you?

3. What are the differences between what we give you and what you need?

Once you have a clearer idea about your area's contribution to your internal customer, you then have a better definition of the value you add in providing service quality. This idea of adding value is the measure of contribution, rather than simply describing what your area does. For example, training units create and provide training programs (this is what they do), but their efforts contribute more skilled employees to the organization (this is their added value).

SOME SOLUTIONS FOR CONSIDERATION

This group of questions focuses on using customer expectations to design your service. Many negative responses here signal a lack of management commitment to service quality. Managers don't see that their leadership in this area is crucial to the success of a service quality strategy.

Table 8.2

Is the Customer's Viewpoint Reflected in Our Service's Design and Standards?	Yes	No
Do you have effectiveness measures in place for all your major service transactions?		
Do the above measures reflect what customers *really* want?		
Have you communicated these effectiveness measures to all employees? Have they accepted them?		
If your effectiveness measures have been met consistently over time, have you made them more challenging?		
Have you set process goals as well as outcome goals?		
Is there a process in place to help you translate customer wants and expectations into service goals?		
Do you believe that you can meet customers' expectations given your current organization and work force?		
Have you communicated your service goals to all employees in your organization?		
Have you translated broad service goals into job-specific goals for each employee?		
Is service commitment to the customer rewarded in your organization?		
Are resources made available to fund service quality goals?		
Have you analyzed your customer-provider work flow to identify where the process is breaking down?		
If the design of your service delivery is faulty, is your top management committed to changing it?		
Does your organization emphasize service goals equally with sales goals?		

Sometimes, however, managers feel committed but don't realize that this commitment should be expressed in tangible and public ways. Commitment to customers should be expressed consistently and in as many ways as managers can think of. If the organization does not have a statement of its values, managers should compose one—the simpler the better. This could take the form of a mission statement, a vision, guiding principles, or beliefs.

Once this statement is established, managers can and should refer to it in all their communications, internal or external. Speeches delivered to employees or customers should always refer to the service quality values of the organization. Events and celebrations can be staged around these values and can include employees or customers. Managers should gear their own conduct and personal and professional standards to support and be consistent with the values of service quality and customer orientation. Employees will know that the achievement of service quality is the driving force of an organization only when the actions of managers prove they are serious about it—when service quality becomes more than a slogan.

It follows that if the organization's values are explicit and communicated consistently, they should also form the foundation of the service quality goals managers set. Service goals become the next layer in the internal translation of customer needs and expectations, following the statement of values.

Ideally, service goals remain consistent with what is important to the customer and, at the same time, are meaningful to people inside the organization. They should also be clear and communicated in such a way that everyone understands how they support the larger mission. Service goals, like goals for sales, profits, expense levels, market share, and so forth, should also be measurable. They should be held in equal esteem with goals mentioned for any of the preceding indicators of success or any others more appropriate for your organization.

Sometimes, managers feel defeated by customers' seemingly "impossible" requests. Top managers should create an atmosphere that encourages a positive attitude—one that fosters opportunities for creative or innovative responses to customers. If managers don't believe it's possible to respond to customers' needs and expectations, they will stay in that rut unless they have a good understanding of the current system and feel supported to take the risks associated with innovation. This also means reducing fear of failure, a sickness common in today's organizations.

Creating an atmosphere that encourages innovation also leads to using technology to enhance service delivery. This can happen in two ways. Technology can help make a service predictable for customers, consistently meeting expectations no matter where or when they want the service. A good example of this is the automatic teller machine. This device makes it possible for a customer to conduct simple business with a bank at his or her convenience, not the bank's.

Another way technology can support service quality is by making it easier for customer contact employees to serve customers. This includes the use of computers. An example of this is the creation and integration of customer databases, so that when a customer asks a question or makes a request, the customer contact person can respond instantly. Other technology may seem less sophisticated, but it is just as important; for example, the installation of a new form letter system that allows easier customizing of letters to customers.

Table 8.3

Can Your Organization Really Provide Excellent Service to the Customer?	Yes	No
Do employees and managers know the organization's service goals?		
Do your employees clearly know what is expected of them on the job?		
Have you identified the service quality behaviors on which employees will be evaluated?		
Are employees rewarded for putting customers first, even if it means disrupting internal procedures?		
Are service criteria included in your job qualifications when hiring new employees?		
Have you allowed for flexibility in the enforcement of your administrative procedures?		
Can your customer contact employees handle exceptions themselves? 　When they must refer exceptions, must they go: 　• to someone of a higher level in their own unit? 　• to their manager? 　• to their manager's boss? 　• to another unit?		
Are managers fully informed about the training given to their employees? Were they consulted when it was designed?		
Are your employees trained on all aspects of their job so that they can be relied on to serve customers? 　Are they trained in: 　• product and service knowledge? 　• systems and procedures? 　• courtesy/customer contact skills? 　• basic writing and/or speaking skills? 　• interpersonal skills? 　• use of essential equipment, including telephone systems? 　• customer needs, expectations, and problems? 　• stress management techniques?		

Table 8.4

Can Your Organization Really Provide Excellent Service to the Customer?	Yes	No
Are your employees evaluated on service as well as productivity goals?		
Are your employees given the opportunity to participate on teams?		
Do you make use of cross-functional teams for solving service problems?		
Is there cooperation among the managers and employees from different areas of your organization?		
Do your employees understand and act on the concept of internal customers?		
Do you know if employees are satisfied with the way they are treated in your organization? • Do they believe they are fairly paid? • Do they beleive they have access to advancement? • Do they feel adequately trained to do their jobs? • Do they feel they are valued members of the organization?		
Are employees at all levels of the organization asked to contribute to decisions that affect their work?		
Does your organization provide a way for employees to discuss new ideas? Are they encouraged by managers to do so?		
Are employees aware of the impact they make on the external customer, no matter where they are in the customer-provider work flow?		
Do your customer contact employees feel they have the ability to provide the level of service customers expect?		
Do you provide customer contact employees with "down time" away from customers?		

MEETING THE DELIVERY CHALLENGE

Just as there are many contributing factors to creating quality customer service, there are just as many ways to reduce it. First is the communication of the service goals. It is hard for employees to believe that organizations can seriously build and demand better teamwork without first making sure that everyone understands the common goals they are working toward. Mixed messages about goals will only lead to cynicism and lower morale. For example, if top management communicates a vision that includes complete customer satisfaction, but unit goals call for reducing needed staff, elimination of skills training, or the introduction of inflexible, complex rules for dealing with customers, then it is no wonder that employees feel the stress of conflict.

Delivering clear communications extends to individual job expectations. Employees need to be clear exactly what management expects of them on the job. This may sound obvious, but research in this area over a period of many years has revealed that this is an area fraught with problems. One way to make sure this is happening is to include a "roles and responsibilities" exercise in training for a particular job. This exercise asks employees to complete a list of what they understand are their job tasks, outcomes, responsibilities, and authorities. They also rank the tasks and outcomes in order of importance and, furthermore, what percentage of time they spend on each. The manager to whom they report directly also completes a similar list, with rankings. Then, the two come together to compare lists. This provides an opportunity for an open discussion and negotiation that leads to clarification on both parts of what is expected on the job.

Employee selection is another area of concern for service providers. Very simply, when the right people are matched to the right jobs, there is less of a gap in providing service. Typical service organizations place the lowest-paid employees in the most sensitive customer contact jobs. Also, they typically provide little training, believing (wrongly) that these jobs require very little skill.

Creation of a carefully designed selection process can go a long way to ensure service quality, including setting educational levels, deciding where to recruit, and developing job-specific evaluations. If possible, include realistic service situations in the evaluation process, asking job candidates to respond as they would on the job. This will provide you with a realistic view of the candidate's skills, while providing candidates with a more realistic view of what the job is all about.

Another idea for strengthening your selection process is to make candidates partners in selection. Allow candidates to spend time, perhaps an hour or more, directly in units providing service. For example, if it is a hotline phone team, allow them to sit on the team listening to customer phone calls (be careful about confidentiality issues). Let them experience real service providers responding to customers. If your candidates cannot visit the units where they will be working and if you have the resources, video can be helpful. Showing candidates what the job is like on tape allows them a degree of self-selection that is not otherwise possible.

Selection and hiring practices can be looked at in the same way as preventive medicine—the time, money, and effort used for this activity prevent serious employee problems later.

Important reasons qualified candidates will choose to work for your organization are the career and growth opportunities within. You can ensure this kind of growth by designing jobs so that everyone is given "whole jobs"—jobs that provide opportunities for the use of a broad range of skills to produce meaningful outputs. This can be accomplished by cross-training, which doesn't mean that everybody knows all jobs but that they can step into a job with minimal effort.

Other ways you can make your organization attractive are as follows:

1. Supporting employees who engage in outside educational activities

2. Limiting the time period that they will be on any one customer contact job

3. Promoting from within the organization and including customer contact employees for consideration even if it involves retraining

Training is a key issue in providing service quality. Nothing substitutes for employees who know their jobs well in the delivery of quality service. Of course, they should be made aware of how their skills support the organization's goals. But nothing affects the dimensions of service quality described in Chapter 4 as much as the knowledge and skills of employees.

Thus, training becomes a high priority to any organization wishing to serve customers well. No dimension of the job should be left out. They should receive training in the following areas:

1. Product and service features in enough detail to foster confidence in discussing them with customers

2. Procedures and methods—not just to do them well under ordinary circumstances but to know them with the depth necessary to change them in extraordinary circumstances

3. Use of electronic systems; familiarity with both hardware and software

4. Knowledge of the customer-provider work flow—they should be able to tell you where their work comes from, where it goes, what value it adds, and so on

5. Knowledge of any techniques useful in quality analysis, such as those mentioned in Chapter 3

6. Enough knowledge of customers so they know not only what managers expect but what customers expect from them

7. Interpersonal skills, including effective behavior in team meetings

Training should be viewed as another opportunity to refresh employees about the overall vision of the organization. Employees should be made aware that their managers are inviting them to training and that they take seriously the time spent away from work. Managers should be involved in training from the beginning, even before an employee goes to class, and should be consulted as to the content of what is presented—and may even be recruited to deliver some training themselves. They should always follow up with employees on all aspects of training, making sure they learned and that there is a transfer of skills to the workplace.

Just as training employees is important, so too is the training of managers and supervisors in the skills needed to do their jobs. Training in the organization's managerial processes, such as planning and budgeting, is essential, as is training in any quality tool and technique used throughout the organization. Also important is training in leadership, as it is needed in a service organization: being able to communicate a vision of service quality, creating goals that support it, creating an environment in which employees feel confident enough to offer ideas; motivating others by setting high levels of personal conduct and effectiveness; and demonstrating a commitment to excellent service to customers.

The previous list may appear to be a tall order for any manager to achieve, but in a service organization, achieving personal quality ought to be everyone's goal because of the very nature of the service relationship. When an employee is serving a customer, at that moment, that employee represents the entire company to that customer—and no one is off the hook.

An excellent way to motivate employees to seek personal quality is to involve employees in the decisions that affect their work. In our experience, nothing motivates people more than the feeling that they contribute and have some control over their work. Teamwork accomplishes this by facilitating communications and cooperation toward achieving common goals. Focus on teams in whatever way works in your organization. They can be formed at all levels; they can include managers and employees, customers, and suppliers. In units where there is heavy production, rotating employees on teams helps to alleviate the tension between group participation and individual production. We have seen quality councils, natural work teams, user groups, partnership boards, and service improvement teams. It doesn't matter what you call them, as long as they follow guidelines similar to those discussed in Chapter 6.

When starting teams, remember to keep them focused. Provide both process and content leadership; allow the team to set some of their own ground rules. As mentioned in Chapter 6, give them a common language, such as a problem-solving method, and make them as cross-functional as possible. Be clear as to your expectations of the team. If you expect them to really make changes and implement solutions, give them the authority to do so. Also, remember to explain (over and over) that being on a team is not something "extra" an individual does and then gets back to his or her real job. Make sure that employees feel supported enough that being on teams does not become an added burden of work.

You may eventually find that your teams more accurately portray the way you do business than does your formal organizational hierarchy and structure. Some organizations we have known have eventually changed the formal structure to reflect team composition.

Employees can also make decisions that affect their work by being delegated the authority to do so; in other words, empowering them. This can be done by reducing the rules and regulations under which they must perform their jobs. It means that they are provided enough training to be confident in any situation with customers. It means that you provide them with customer "scripts" for difficult customer situations, including nonroutine requests or setting up procedures for unusual circumstances. It could also mean that the organization risks communicating overall guidelines, such as "Whatever an employee does to satisfy a customer, if not illegal or immoral, will be supported by management."

When managing a service organization, you need to think carefully about recognition and rewards. If service is truly valued equally with sales, for example, then you must make sure that extraordinary service is rewarded just as great sales performance is rewarded. Otherwise, employees will perceive that there is a mixed message and conclude that there is a lack of commitment to service quality.

There are many ways to recognize employees in nonmonetary ways that are appreciated; in fact, in our experience, nonmonetary rewards are more remembered and linked to performance than is money given outright. This doesn't mean only trophies or plaques, although when linked to celebrations or events, these symbols can be powerful. It can also mean gifts, such as household items, paid dinners, or an extra day off. Only the limits of your imagination stand in the way of inventing ways to recognize employees in nonmonetary ways.

On the other hand, monetary rewards linked to service goals remain powerful incentives to repeating wanted behaviors. If a formal performance appraisal system exists in your organization, make sure that the standards reflected are closely tied to service goals and that the rewards that flow from it reflect those goals as well.

Individuals need not be the only ones rewarded for attaining service goals. When teams perform in outstanding ways, they, too, should receive recognition. This recognition can be monetary, such as when the team shares in a percentage of savings that result from implementation of their solutions. It can also be nonmonetary, such as teams competing for prizes (trophies, scrolls, or banners) in an annual event held expressly to honor teamwork. Remember to recognize not only teams that achieve what is easily measured (such as savings in money, time, staffing, or space) but teams that achieve what may be harder to measure, such as increased customer satisfaction.

Finally, if you do not do so already, and your employee population is large enough, you may wish to establish employee surveys, just as you do for customers. You should do a survey that treats employees as if they were customers of your human resources policy and ask questions about compensation, benefits, fairness, career growth, quality of leadership, the quality of training, and so on. Just as with customers, you can create your own or purchase one of the numerous generic surveys currently on the market. Just as anonymity works well for customers, it also works to get information about your employees you may not otherwise get. You can also include questions that refer specifically to how they are treated as internal customers, their perception of the organization's commitment to customers, and the success of service quality efforts.

SOME IDEAS ABOUT INTERNAL COORDINATION

Many organizations find themselves in the position of doing good things for the customer but never highlighting that to them. The opposite is also true—communications

Table 8.5

Is the Customer's Viewpoint Reflected in Our Service's Design and Standards?	Yes	No
Do your customers receive consistent service at all locations (branches, units, stations, or outlets)?		
Are procedures in place to ensure consistent service?		
Do you have a method for sharing and implementing new ideas from one location to another?		
Does your human resources unit treat other employees as internal customers?		
Do the people who work in your operations or production units ever meet face to face with external customers?		
Are sales employees acknowledged as internal customers by other employees (production, staff, operations, etc)?		
Do marketing people ever discuss their plans with: • operations or production units? • other support units, such as information systems? • customer contact employees or managers?		
Are your operations managers aware of the content of sales materials prepared for prospective customers?		
Do your operations managers have the opportunity to discuss and modify the content of sales materials *before* they are used by sales employees?		
Does marketing inform customer contact employees about changes in products or services *before* telling external customers?		
Does marketing inform customer contact employees about the introduction of new products or services *before* telling external customers?		
Does sales training in your organization include a thorough overview of operations and service support?		
Are your sales employees discouraged from promising more than the organization is capable of delivering?		

about the service are so misleading that they have an adverse effect on either the customer's expectations or perceptions of service performance.

A customer that has come to expect certain levels of service from you in one local area will expect similar levels wherever they see your outlets. In a sense, the act of providing service establishes future customer expectations. It "advertises" that what customers are experiencing during the service delivery is what the organization is capable of providing each and every time.

One of the ways to ensure positive customer perceptions of your organization is to provide consistent service no matter where or when the customer demands it. Ensuring consistency of delivery includes implementing the most efficient processes in local areas, whether you have outlets, units, counters, or stations dispersed geographically. Managers should attempt to reproduce process enhancements and efficiencies across the organization by sharing new ideas and problem solutions. Of course, this assumes that the organization has a method in place to capture and share ideas. Remember that service excellence demands consistency of delivery, not rigid adherence to inflexible rules.

It also assumes that you have thought about how you are going to ensure service consistency across the organization. Systems, methods, and procedures must be quality assured through auditing measures, whether this is done by your own employees or from outside. Actually, it is reasonable to use both methods.

Perhaps the hardest job is to ensure what is the most basic feature of service delivery: one provider serving one customer. Unlike systems, processes, and certain technologies, humans are noted for their individuality. The challenge, then, becomes how to ensure the quality in a personal aspect of the service relationship. Some suggestions follow:

1. Training can help reduce the worst variations among employees but is not meant to produce robots (nor should it).

2. Customer "scripts" also help reduce variations in unusual circumstances.

3. If customer contact employees are on the telephone, call monitoring helps ensure consistency.

4. Use of techniques, such as "mystery shopper" or "mystery guest," helps ensure such individual features as courtesy or friendliness.

Problems also develop when the various functions in the your organization do not coordinate their activities to support one another. This includes efforts to coordinate the goals and activities of the different functions in your organization: human resources, marketing, sales, operations, and information systems. They should all know and understand common organizational goals and produce plans that support them.

One way to overcome this difficulty in coordination is to develop an integrated planning approach so that all functions support each other. A planning approach of this kind involves as many people in the organization as is necessary to produce a coherent plan; planning is not left to a special staff or planning "experts." The plan need not be complex but must include coordinated efforts between functions and levels.

Let's face it. Sometimes service providers are their own worst enemies when communications to customers are too optimistic about what the organization can really provide. Promising too much is a mismanagement of customer expectations and leads to disappointment. Make sure that advertising, in whatever form, accurately reflects your current capabilities and that sales employees are truly educated about the company, including the importance of other functions, such as operations and service support staff. In fact, it would be helpful if sales employees spend time in operations or customer service units, seeing firsthand how the organization works.

Marketing should also seek the advice and input of other functions, including operations and staff, when designing and launching its campaigns. There should be a partnership formed among these areas to achieve the common goal of excellent service. Furthermore, staff support, operations, production, and customer service units should be told in advance of new advertising claims, guarantees, or service promises—and are allowed to discuss and negotiate changes to the content. In fact, it is a good idea for marketing to go directly to customer contact employees for advertising ideas.

SUMMARY

Following the recipe for service quality described in this book will earn you loyal customers and steady growth—and a reputation for service excellence. No one loses with this type of strategy. Managers win, stockholders win, employees win, and certainly customers win. And because we are customers of some service every day, if excellent service quality becomes the standard to which all organizations aspire, each of us will win every day.

B O O K S

Albrecht, Karl. *Organization Development: A Total Systems Approach to Positive Change in Any Business Organization*. Englewood Cliffs, NJ: Prentice-Hall, Inc., 1983.

Albrecht, Karl, and Lawrence J. Bradford. *The Service Advantage: How to Identify and Fulfill Customer Needs*. Homewook, IL: Dow Jones-Irwin, 1990.

Argyris, C. *Increasing Leadership Effectiveness*. Melbourne, FL: Krieger Publishing, 1983.

Aubrey, Charles A., and Patricia K. Felkins. *Teamwork: Involving People in Quality and Productivity Improvement*. Milwaukee, WI: Quality Press, 1988.

Barker, Joel Arthur. *Discovering the Future: The Business of Paradigms*. St. Paul, MN: ILI Press, 1985.

Blake, Robert R., and Jane Srygley Mouton. *The New Managerial Grid*. Houston, TX: Gulf Publishing Co., 1983.

Block, Peter. *Flawless Consulting: A Guide to Getting Your Expertise Used*. San Diego, CA: University Associates, Inc., 1981.

Crosby, Philip B. *Quality Is Free: The Art of Making Quality Certain*. New York, NY: McGraw-Hill Book Co., 1979.

Deal, Terrence E., and Allan A. Kennedy. *Corporate Cultures: The Rites and Rituals of Corporate Life*. Reading, MA: Addison-Wesley Publishing Co., 1982.

Doyle, Michael, and David Strauss. *How to Make Meetings Work*. New York, NY: The Berkley Publishing Group, 1976.

French, Wendell L., and Cecil H. Bell, Jr. *Organization Development: Behavioral Interventions for Organization Improvement*, 2nd ed. Englewood Cliffs, NJ: Prentice-Hall, Inc., 1973.

Horovitz, Jacques. *Winning Ways: Achieving Zero-Defect Service.* Cambridge, MA: Productivity Press, 1990.

Juran, J. M. *Quality Control Handbook,* 3rd ed. New York, NY: McGraw-Hill Book Co., 1974.

Miller, Lawrence M. *American Spirit: Visions of a New Corporate Culture.* New York, NY: William Morrow & Co., Inc., 1984.

Pascarella, Perry. *The New Achievers: Creating A Modern Work Ethic.* New York, NY: The Free Press, 1984.

Peters, Tom, and Nancy Austin. *A Passion for Excellence: The Leadership Difference.* New York, NY: Random House, 1985.

Prince, George M. *The Practice of Creativity: A Manual for Dynamic Group Problem Solving.* New York, NY: Harper & Row, 1970.

Steele, Fritz. *The Role of the Internal Consultant: Effective Role-Shaping for Staff Positions.* Boston, MA: CBI Publishing Co., Inc., 1982.

Weisbord, Marvin R. *Organizational Diagnosis: A Workbook of Theory and Practice.* Reading, PA: Addison-Wesley Publishing Co., 1978.

Weisbord, Marvin R. *Productive Workplaces: Organizing and Managing for Dignity, Meaning and Community.* San Francisco, CA: Jossey-Bass, Inc., 1987.

Zeithaml, Valarie A., A. Parasuraman, and Leonard L. Berry. *Delivering Quality Service: Balancing Customer Perceptions and Expectations.* New York, NY: The Free Press, 1990.

Berry, Leonard L., Valarie A. Zeithaml, and A. Parasuraman. "Five Imperatives for Improving Service Quality," *Sloan Management Review,* Summer 1990, pp. 29–37.

Berry, Leonard L. "Service Excellence for Fun and Profit," *Total Quality Management,* The Conference Board Report No. 963, pp. 39–41.

Jacobson, Allen F. "Commitment to Quality: Relentless, at All Levels," *Financier,* May 1990, pp. 37–40.

Michaelson, Gerald A. "The Turning Point of the Quality Revolution," *Across the Board,* December 1990, pp. 40–45.

Oberle, Joseph. "Quality Gurus: The Men and Their Message," *Training,* January 1990, pp. 47–52.

Parasuraman, A., Valarie A. Zeithaml, and Leonard L. Berry, "A Conceptual Model of Service Quality and Its Implications for Future Research," *Journal of Marketing,* Vol. 49 (Fall 1985), pp. 41–50.

Parasuraman, A., Valarie A. Zeithaml, and Leonard L. Berry. "SERVQUAL: A Multiple-Item Scale for Measuring Customer Perceptions of Service Quality," Report No. 86–108, Cambridge, MA: Marketing Science Institute, 1986.

Reichheld, Fredrick F., and W. Earl Sasser, Jr. "Zero Defections: Quality Comes to Services," *Harvard Business Review,* September-October 1990, pp. 105–111.

Wood, Robert Chapman. "The Prophets of Quality," *The Quality Review,* Winter 1990, pp. 18–25.

Zeithaml, Valarie A., Leonard L. Berry, and A. Parasuraman, "Communication and Control Processes in the Delivery of Service Quality," Report No. 87–100, Cambridge, MA: Marketing Science Institute, 1987.